Andrew Robinson is the author of more than twenty-five books on science, the history of science and the arts, including the award-winning *Earthshock*; *The Story of Measurement*; and *Genius: A Very Short Introduction*. He has written biographies of Albert Einstein (*A Hundred Years of Relativity* and *Einstein on the Run*) and Thomas Young (*The Last Man Who Knew Everything*). He holds degrees from Oxford University (in chemistry) and the School of Oriental and African Studies, London, and was a visiting fellow of Wolfson College, Cambridge, from 2006 to 2010. Having been literary editor of the *Times Higher Education Supplement* for twelve years, in 2007 he became a full-time writer and journalist. He reviews regularly for newspapers and magazines, including the science journals *Nature*, *Science* and *The Lancet*.

THE
SCIENTISTS

PIONEERS
of
DISCOVERY

Edited by
ANDREW ROBINSON

with 27 illustrations

On the cover: *see* Sources of Illustrations

First published in the United Kingdom in hardcover in 2012 by
Thames & Hudson Ltd, 181A High Holborn, London WC1V 7QX

This compact paperback edition published in 2023

The Scientists © 2012 and 2023 Thames & Hudson Ltd, London

Text by Andrew Robinson

British Library Cataloguing-in-Publication Data
A catalogue record for this book is available from the British Library

ISBN 978-0-500-29706-3

Printed and bound in the UK by CPI (UK) Ltd

MIX
Paper | Supporting
responsible forestry
FSC® C171272
www.fsc.org

CONTENTS

INTRODUCTION
On the shoulders of giants

*If I have seen further it is by standing
on the shoulders of giants.*

ISAAC NEWTON IN A LETTER TO ROBERT HOOKE, 1676,
CITING THE FRENCH MEDIEVAL PHILOSOPHER BERNARD
OF CHARTRES (D. AROUND 1130)

In ancient Rome, the Latin word *scientia* – from which *science* first emerged in Old French during the Middle Ages – meant 'knowledge' in the widest possible sense. Originally, the term encompassed not merely knowledge obtained from studying nature, but every intellectual discipline, including philosophy, politics and theology. Even today, it is common to find phrases such as the 'science of logic', 'political science' and the 'dismal science' (in reference to economics).

In this book, however, 'science' carries its general modern meaning: the collection, classification and analysis of data about the physical universe. This kind of science is as old as the Babylonians, Egyptians and Greeks of the ancient world, not to mention the medieval Arabs, Chinese, Europeans, Indians and Persians of the millennium ending around 1500, whose scientific contributions are increasingly recognized by historians. Yet surprisingly, the modern word to describe those doing such science, 'scientists', is less than two hundred years old. Before the mid-19th century, individuals we would now routinely call great scientists, such as the 17th-century figures Johannes Kepler, Galileo Galilei and Isaac Newton, were instead known as 'natural philosophers'. As late as 1799, when the progressively minded Royal Institution opened its doors in London, it appointed a 'professor of natural philosophy' to lecture on what we would now call physics.

The birth of a profession

'Scientist' was coined only in 1833, at a meeting of the newly founded British Association for the Advancement of Science in Cambridge. One of the attendees, the poet Samuel Taylor Coleridge, had raised the question of what name to give to the fast-increasing tribe of professional experts in various scientific disciplines: an umbrella term that would cover anatomists, astronomers, biologists, botanists, chemists, geologists, palaeontologists, physicists, zoologists and others. According to the report of the ensuing discussion, 'philosopher' was rejected as 'too wide and too lofty a term', whereas 'savant' was regarded as 'rather assuming' – and also too specifically French; and so the association's polymathic secretary, William Whewell, a mineralogist, historian of science and future master of Trinity College – where Newton had once studied – proposed 'scientist', by analogy with 'artist'. Although the new name was 'not generally palatable' to the meeting, it immediately caught on in the United States, was gradually adopted in Britain, and was universally used by the first half of the 20th century, during which science and technology came to dominate, and even to define, the modern world.

Notwithstanding science's new prestige, 'The whole of science is nothing more than a refinement of everyday thinking', claimed Albert Einstein in 1936, when he was already regarded as the most original scientist since Newton. This was a mischievous paradox typical of a man with a genius for discovering simplicity in complexity. True if you were Einstein, maybe – but for most of us it is a scarcely credible statement. Come off it, one cannot help thinking. What have our everyday thought processes got to do with the thinking of great scientists like Einstein, especially the esoteric mathematical subtleties of 20th-century physicists?

Physics, in its long evolution towards unifying more and more of the physical universe on the basis of fewer and fewer fundamental ideas, seems to have moved further and further away from everyday thinking with each passing decade. Most non-physicists learn to use some of the technological by-products of pure research in physics: computers, mobile phones, the World Wide Web, and the like. But general relativity – Einstein's theory that explains black holes and the accuracy of the satellite GPS – and quantum theory – which underlies plasma televisions and lasers – appear to have nothing in common with everyday experience. Earlier key scientific ideas, such as Archimedes' principle of displacement and flotation, Newton's laws of motion and gravitation, Michael Faraday's concept of the magnetic field and Charles Darwin's principle of evolution by natural selection, are comparatively accessible to everyday thinking; we can even do simple experiments at home to demonstrate the truth of them with objects immersed in water, falling coins, patterns of iron filings and moving compass needles. Not so with relativity and quantum theory.

The influence of the ancients

Modern science owes much, of course, to ancient Greeks of the calibre of Archimedes, Aristotle, Democritus, Euclid, Eratosthenes and Ptolemy. To these mathematicians and natural philosophers who observed nature and thought for themselves two millennia and more ago, we owe, for example, the earliest systematic classification of animals, the perception that matter is made of atoms, the invention of geometry, the idea that light travels in straight lines, the first estimate of the Earth's circumference and the concept of latitude and longitude. Here the ancients' thinking was marvellously fruitful.

However, the ancient Greeks also (with one or two enlightened exceptions such as Aristarchus of Samos) believed that the Sun and the planets revolved around the Earth in perfect circles, and that the more massive a body was, the faster it would fall when dropped. Aristotle's 'everyday thinking', it would seem, led him to the conclusion in his *Mechanics* that, 'The moving body comes to a standstill when the force which pushes it along can no longer so act to push it' – a grossly inaccurate conception of mass and force. A more massive body should fall faster because, said Aristotle, it had a greater tendency to seek the centre of the Earth – which is also simple to demonstrate as wrong. His concept of 'motion' included not only pushing and pulling, but also combining and separating and waxing and waning. A fish swimming and an apple falling from a tree were, obviously enough, in motion – but so too were a child growing into an adult and a fruit ripening. Thus, common sense led Aristotle, who was not a great experimenter (unlike Archimedes), into a hopeless conceptual muddle about the simplest facts of mechanics.

Yet such was the prestige of the Greeks in philosophy that Aristotle's ideas about the physical world dominated European intellectual life right up to the time of Newton in the 17th century, and even to the time of Darwin, who deeply admired Aristotle's observations on animals. In the 1620s, the English natural philosopher Francis Bacon, whose writings were soon to be instrumental in the launch in 1660 of the world's oldest surviving scientific society – London's Royal Society – remarked critically of his predecessors that, 'All the philosophy of nature which is now received, is either the philosophy of the Grecians, or that other of the alchemists.... The one is gathered out of a few vulgar observations, and the other out of a few experiments of a furnace. The one never faileth to multiply words, and the other ever faileth to multiply gold.'

Intellect and observation

But challenges to Aristotle's view of the world were underway. In 1543, on his deathbed, Nicolaus Copernicus published *De revolutionibus orbium coelestium*, his heliocentric picture of the Solar System with the Earth and other planets revolving around the Sun. This challenge to both everyday perception and biblical scripture was greeted with much scepticism by natural philosophers and, in due course, resistance from the Catholic Church – but eventually, the grand idea caught on that the Earth was no longer at the centre of the world, as it had been since the time of the Greeks, although Copernicus still adhered to the ancient view that the planetary orbits were circular.

Then, in 1609, Kepler, using the first accurate observational data on the planets' movements, compiled by Tycho Brahe, made an educated guess that their orbits around the Sun were not circles but ellipses – one of the geometrical forms discovered by the ancient Greeks – and with this mental leap Kepler conceived his laws of planetary motion. These enabled him to calculate astronomical tables, and hence the positions of the planets at any time in the past, present or future, which fitted well with astronomers' observations. As Einstein remarked on the tercentenary of Kepler's death in 1930, 'Kepler's marvellous achievement is a particularly fine example of the truth that knowledge cannot spring from experience alone but only from the comparison of the inventions of the intellect with observed fact.'

Around the same time as Kepler, Galileo Galilei overturned Aristotle's erroneous notions of motion through physical, quantitative experiments with moving objects and falling weights. He showed that a body moving at a constant speed – that is, uniformly – does not require to be 'pushed' by a force, as Aristotle had claimed. For example, a marble set rolling at a given speed on a perfectly horizontal, frictionless floor will continue to move at that speed. (In the real world, the force of friction will eventually bring it to a stop.) And he demonstrated that the speed of a freely falling body does not depend on its mass. Hence the cannonballs of different mass that he allegedly dropped from the Leaning Tower of Pisa at the same instant were found to hit the ground at the same time, not at different times (as Aristotle would have predicted). 'Pure logical thinking cannot yield us any knowledge of the empirical world; all knowledge of reality starts from experience and ends in it. Propositions arrived at by purely logical means are completely empty as regards reality', wrote

Einstein some three centuries later. 'Because Galileo saw this, and particularly because he drummed it into the scientific world, he is the father of modern physics – indeed, of modern science altogether.'

'Discoverers of undiscovered things'

Mainly for this reason, *The Scientists* begins with articles on Copernicus, Kepler and Galileo – the progenitors of the Scientific Revolution of the 16th and 17th centuries that undoubtedly laid the foundations for present-day science. However, the rest of the book does not follow a purely chronological progression, century by century, preferring instead to emphasize the development of scientific thinking over time in particular fields such as cosmology. Thus, the six sections start with science at the largest scale – that of the Universe – and then progressively reduce the scale, to that of the Earth (section two), of molecules and matter (section three) and of the subatomic world (section four). Sections five and six are devoted to the living world of plants and animals, including of course the human body and mind. Although all of the core scientific disciplines are covered along the way, including psychology, other disciplines with a scientific element, such as mathematics, medicine, archaeology and anthropology, have been omitted so as to keep the book to a manageable size, with the exception of three articles on the physicians-cum-physiologists William Harvey and Jan IngenHousz, and the archaeologists-cum-anthropologists of human origins Louis and Mary Leakey. Also omitted, again for lack of space, are applied scientific thinkers such as Christopher Wren, James Watt and Thomas Edison, whose chief achievements lay in architecture, engineering, technology or invention.

Every article mingles a scientist's life with his or her science. For all the supposed objectivity of the scientific method, the progress of science has always been driven by strong personalities. The biographies (and autobiographies) of great scientists often illuminate their motivations and sometimes offer clues as to the sources of their breakthroughs. Who can resist the anecdote of Archimedes, the king's gold crown and the overflowing bathtub? Or Newton's reported comment about gravity and the falling apple? Or Einstein's own description of his teenaged thought experiment about chasing a light ray? Or James Watson's rumbustious account of the decoding of the structure of DNA in *The Double Helix* – a book at first condemned by Watson's collaborator Francis Crick for its personalizing of scientific research, but in due course grudgingly admired by him for its truthfulness.

Apart from their love of science, do the forty or so great scientists chosen here share anything in common? Their nationalities, family backgrounds, education and training, personalities, religious beliefs, working methods and the circumstances of their greatest discoveries differ enormously. But in one respect, at least, they do appear to be alike: all of them worked habitually and continually at science and were prolifically productive. The French mathematician, theoretical physicist, engineer and philosopher of science Henri Poincaré (a seminal influence on Einstein's theory of special relativity) published 500 papers and 30 books; Einstein himself produced 240 publications; Sigmund Freud had 330. As Darwin observed to his son late in life, 'I have been speculating last night what makes a man a discoverer of undiscovered things, and a most perplexing problem it is. Many men who are very clever – much cleverer than the discoverers – never originate anything. As far as I can conjecture, the art consists in habitually searching for causes or meaning of everything which occurs. This implies sharp observation and requires as much knowledge as possible of the subject investigated.' Newton, when asked how he had discovered the law of gravity, replied pithily, 'By thinking on it continually.' All of the great scientists in this book would, one senses, have said amen to that.

UNIVERSE

From earliest times, the vastness of the Universe could be sensed, if not comprehended, simply by looking upwards at the night sky: at the Moon, the Sun, the moving planets, occasional comets and the fixed stars of varying brightness. The forces holding these heavenly bodies in their positions in space were a mystery that preoccupied thinkers from the ancient Greeks up to 20th-century astronomers like Edwin Hubble, who discovered that the Universe is not in a steady state but expanding from a moment of creation, subsequently termed the Big Bang. For scientists living before the 19th century, including Copernicus, Kepler, Galileo and Newton, God was the obvious ultimate explanation of these forces – their 'first cause'. But how, they asked, did God's laws actually operate in the physical world of space, matter and time?

Aristotle adhered to a mechanical philosophy. He believed that a force could produce motion through space only by physical contact with an object. He therefore rejected the concept of empty space, the vacuum, partly on the grounds that space had to be filled with a substance to transmit the force acting between the Earth and the Sun. Kepler agreed and postulated that the invisible solar force was magnetic, although he could not explain magnetism. So did René Descartes, writing soon after Kepler, who envisioned space to be full of invisibly small particles of matter moving always in circular streams or vortices, which pushed visible matter, including the heavenly bodies, like a wind. Earth lay at the middle of a small vortex, said Descartes, which created gravitational attraction towards its centre and held the Moon in its orbit.

The young Newton sympathized with Cartesian vortices, yet his eventual mathematical theory of gravity – with its famous inverse-square law relating the attraction between masses to their distance apart – did not fit well with the mechanical philosophy. For all its excellence at predicting motion, it offered no mechanism for gravity's apparently instantaneous 'action at a distance', as Newton was acutely aware. In *Principia mathematica*, he justified his theory as follows: 'It is enough that gravity really exists and acts according to the laws that we have set forth and is sufficient to explain all the motions of the heavenly bodies and of our sea.'

Nor could Newton's concept of space and matter provide a satisfactory explanation of light, magnetism and electricity. Newton preferred a mechanical 'corpuscular' theory of light, which treated it as a stream of small particles or 'corpuscules', despite the theory's weakness in explaining colour, reflection, refraction and other optical phenomena. But in the early 19th century, Thomas Young showed that light could behave as a wave by demonstrating that two light beams could interfere with each other. Michael Faraday, and also Lord Kelvin, then showed that electricity and magnetism were interlinked phenomena, which led Faraday to introduce the non-mechanical concept of the continuous 'field' of force – whether electric or magnetic. James Clerk Maxwell used the field concept to reconceive light mathematically as an electromagnetic wave, comprising an electric field and a magnetic field oscillating at right angles to each other. Maxwell's radical theory of electrodynamics was experimentally confirmed by Heinrich Hertz, who showed that light, radio waves and radiant heat were all electromagnetic waves travelling at the speed of light.

Yet, no 19th-century physicist could explain the transmission medium for the vibrating electromagnetic wave. Instead of vortices, Maxwell and others believed that space was made up of an ether. But this had contradictory properties: for various respectable physical reasons, the ether had to be invisible, absolutely stationary, weightless, without the least viscosity, yet stronger than steel and undetectable by any instrument.

Unsurprisingly, the young Einstein was not convinced. Near the beginning of the 20th century, building on Newton's and Maxwell's theories of electro-magnetism, while dismissing the notion of the ether, Einstein created his theory of general relativity. This reconceived the Universe as a curved 'space–time' continuum with a constant speed of light independent of the speed of the observer. So far, his electrodynamical theory of the Universe appears to have withstood extensive experimental tests. Science still lacks, however, a unified theory of gravity and electromagnetism.

Nicolaus Copernicus
Inventor of the solar system
(1473–1543)

In the midst of all dwells the Sun.... Thus indeed,
as if seated on his royal throne, the Sun rules the
family of stars circling round him.

NICOLAUS COPERNICUS, *DE REVOLUTIONIBUS*
ORBIUM COELESTIUM, 1543

As the author of *De revolutionibus orbium coelestium* (On the revolutions of the heavenly spheres), the 16th-century Polish astronomer Nicolaus Copernicus challenged the cosmology of Aristotle and Ptolemy that had reigned since ancient Greece. According to the prevailing Aristotelian-Ptolemaic model of the Universe, a motionless Earth lay at the centre of a sphere of orbiting planets, beyond which lay the stars, which moved around the Earth every twenty-four hours. Turning this theory on its head, Copernicus proposed the idea that Mercury, Venus, Earth, Mars, Jupiter and Saturn all circle the Sun. By doing so, he initiated a revolution that transformed science for ever.

Born in Toruń, Poland, Copernicus was ten years old when his merchant father died. His maternal uncle, Church administrator Lucas Watzenrode, undertook Nicolaus's education, hoping his nephew would follow in his footsteps and become a cleric. In autumn 1491, Lucas, now Bishop of Warmia, enrolled Nicolaus in his alma mater, the University of Kraków, northern Europe's best for astronomical studies. There Copernicus developed a lifelong passion for mathematical astronomy.

Peripatetic scholar

Copernicus began his studies in Kraków with a course in the arts, concentrating on Latin translations of and commentaries on the works of Aristotle. According to Aristotle, Earth and all natural objects are made of four elements – earth, water, air, fire – while the perfect and unchanging heavens are made of a fifth substance, the ether. Aristotle taught that objects in the ethereal sphere move in circular motions. Copernicus also studied geometry, focusing on the work of Euclid and on simplifications of the geometrical astronomical models of Claudius Ptolemy. The Ptolemaic model of planetary motion required two circles. Each planet was assumed to move uniformly along a small circle called an epicycle, which in turn moved along a larger circle around the Earth called the deferent. When the planet's backward motion on the epicycle exceeded the forward speed of the epicycle on the deferent, people on Earth would see the planet moving in retrograde motion – that is, backwards – with respect to the distant stars. The Aristotelian-Ptolemaic model assumed that the planets revolved around Earth, which was fixed at the Universe's centre. Copernicus's revolutionary model would challenge that scheme.

In 1495, after four years in Kraków, Copernicus returned, without taking a degree, to Frombork, a town on the north-east Baltic coast, seat of the Warmia bishopric and site of its cathedral church. When a vacancy occurred in the church's administration, Lucas nominated his nephew. The appointment to the position was contested, but before its finalization in Nicolaus's favour, Lucas sent him to the University of Bologna to study canon law, the rules for governing the Church.

During his time in Bologna, Copernicus became assistant to a famous astronomy professor, Domenico Maria Novara da Ferrara. He had brought with him a printed copy, with his own annotations, of the *Alphonsine Tables*, the tables created for King Alfonso X of Spain in the 13th century that provided data for computing solar, lunar and planetary positions relative to the fixed stars. Copernicus also procured a copy of the 1496 abridgment of Ptolemy's *Almagest* – a mathematical and astronomical treatise proposing the complex motions of the stars and planetary paths – by Regiomontanus, the Latinized name of a German mathematician and astronomer who had died twenty years earlier. Copernicus was troubled by a discrepancy between Ptolemy's planetary orbits and Aristotle's ideals of geometric perfection, which had also troubled both Ptolemy himself and medieval Arab and Jewish astronomers in Spain.

After four years, Copernicus left Bologna, again without a degree. He spent a few months touring Rome, where the 1,500th anniversary of Christianity was being celebrated. What some viewed as Pope Alexander's extravagant expenditure of Church funds for the celebration soon led to the movement for reformation associated with the German monk Martin Luther, which resulted in the breakaway Protestant faith. In 1501, Copernicus petitioned the cathedral chapter in Warmia for two more years of study in Italy, promising to use his medical training in Padua to treat the bishop and members of the chapter. In 1503, he returned to Warmia without a medical degree but with a doctorate in canon law from the University of Ferrara. He now undertook the responsibilities of canon, while also serving seven years as personal secretary and physician to his uncle, who died in 1512.

Modelling the solar system

Copernicus continued to study Regiomontanus's *Almagest* abridgment, discovering that the German had built on Ptolemy's consideration of an alternative to the epicycle to account for the planets' retrograde motions, showing how the roles of the deferent and the epicycle could be converted to an eccentric circle with a movable centre that always lies in the direction of the Sun. This demonstration was an entrée to Copernicus's Sun-centred system.

Copernicus first developed a system in which each outer planet – Mars, Jupiter and Saturn – circled the Sun, while the Sun circled Earth. When he incorporated the inner planets, Mercury and Venus, into the system, Copernicus had to choose between an arrangement of planets circling the Sun, which in turn circled Earth, or a more elegant arrangement, with Earth also orbiting the Sun. The more elegant model would make Earth a planet in motion, thus explaining the retrograde movement of the planets and other seeming irregularities in celestial motion as apparent, not real, and challenging prevailing ideas. Placing all the planets in orbit around the Sun also determined their distances and periods of revolution, which in his model increased together. (The moon is on an epicycle around the Earth.) Decades later, in *De revolutionibus*, Copernicus justified his model: 'In no other arrangement do we find such a sure harmonious connection between the size of the orbit and its period.' He was also concerned about the violation of the principle of uniform circular motion in Ptolemy's planetary theory, the solution to which he found in models of planetary motion developed by earlier Arabic astronomers associated with the observatory of Maragha that preserved uniform circular motion strictly. Historians believe that Copernicus's intellectual breakthrough occurred around 1510, culminating in *Commentariolus* (Little commentary), a small pamphlet on his heliocentric planetary theory, which also preserved uniform circular motion.

In that year, Copernicus settled permanently in Frombork. Much of his official Church work there was financial, leading him to write an essay about currency and coinage. He made time, however, for his main intellectual passion, the observation of planetary and solar positions that would provide the necessary data for revising Ptolemy's models. In 1515, the *Almagest* first appeared in print in its entirety. When Copernicus saw how much more extensive it was than Regiomontanus's abridgment on which his *Commentariolus* was based, he understood the magnitude of his self-imposed task. To convince the Western world to replace the cosmological system it had adhered to for over a thousand years would require decades of observation and computation.

Revolution in the heavens

Copernicus's manuscript grew larger over the years, but despite collegial encouragement, he made no preparations for publication. In 1539, however, a young Austrian mathematician, Georg Joachim Rheticus, came for an extended visit. Despite an edict of 1526 banning Lutherans from Warmia, Copernicus welcomed Rheticus as a disciple. In early 1540, a printer in Gdansk published Rheticus's *Narratio prima* (First report), summarizing for scholars Copernicus's theories. Rheticus worked with Copernicus on final revisions to the manuscript, which in 1542 he delivered to Nuremberg printer Johannes Petreius, who prepared woodblocks for Copernicus's complicated technical diagrams.

During the ten months it took to print the large six-sectioned book, Copernicus sent for inclusion in the preface a 1536 letter from a cardinal urging him 'to communicate your discovery to enthusiasts and to send me at the first possible opportunity your labours on the sphere of the universe together with the tables'. Copernicus also wrote a dedicatory preface to Pope Paul III, attributing his delay in publication to concerns about reception of his ideas, but expressing his sincere belief that they would benefit 'the ecclesiastical Commonwealth'.

In late 1542, a stroke that left Copernicus partially paralysed compromised his ability to work. Only on the day he died, 24 May 1543, did he see a printed copy of his magnum opus. He may have been too ill to notice that an anonymous introduction had been added, calling his arguments 'hypotheses' and claiming his heliocentric model was of use only for mathematical calculation, contradicting

Copernicus's own conviction that the Sun-centred universe was real, not merely useful for computational purposes. In 1609, it was revealed that Andreas Osiander, a Lutheran theologian who had overseen the proofreading on behalf of Rheticus, had inserted this introduction.

Only 500 copies of *De revolutionibus* were printed in 1543, and its radical argument caused no stir. Over the next decades Johannes Kepler and Galileo Galilei went on record as accepting Copernican cosmology as factual, not merely theoretical, but many clerics objected to the conflict between heliocentric and biblical cosmology. In 1616, the Catholic Church put *De revolutionibus* on the Index of Prohibited Books. Within Italy, copies of the book were censored, but elsewhere the ban was ignored. After Isaac Newton's *Principia mathematica* appeared in 1687, more scholars began to embrace Copernicus's system, but it was not until 1835 that his book (along with Kepler's and Galileo's) was removed from the Index.

In recent decades, scientists and Church alike have honoured Copernicus. In 1972, anticipating the 500th anniversary of the astronomer's birthday, NASA launched the *Copernicus* satellite, which studied interstellar matter during its eight years of operation. In 2005, an unmarked grave near Frombork cathedral's altar was shown to contain the remains of a man of about seventy years of age. In 2008, Polish archaeologists reported that DNA comparisons with hair strands in a book known to have been used by Copernicus confirmed the skeleton to be his. On 22 May 2010, Copernicus, who in 1543 lacked the fame to merit a marked grave, was reburied under the cathedral's altar with a black granite tombstone. Also in 2010 the International Union of Pure and Applied Chemistry officially endorsed the name 'copernicium' (Cp), proposed by the scientists who had discovered element 112.

Johannes Kepler

Analyst of planetary motion

(1571–1630)

To God there are, in the whole material world, material
laws, figures and relations of special excellency and of the
most appropriate order.... Those laws are within the grasp
of the human mind; God wanted us to recognize them by
creating us after his own image so that we could share
in his own thoughts.

JOHANNES KEPLER, IN A LETTER TO JOHANN GEORG HERWART
VON HOHENBURG, 9–10 APRIL 1599

Johannes Kepler transformed the ancient tradition of cosmology and laid the groundwork for modern science by treating astronomy – up to then a branch of the liberal arts – as part of mathematical physics. His eponymous laws of planetary motion provided a foundation for Isaac Newton's universal law of gravity, and they still explain the orbits of not only planets, but also dwarf planets, comets, asteroids, trans-Neptunian objects, and even exoplanets, planets orbiting distant stars. His experiments with lenses and mirrors and studies of the human eye also established the field of modern optics.

Kepler was born into a Lutheran family in Weil der Stadt in southern Germany, a free imperial city within the Holy Roman Empire. He earned bachelor's and master's degrees from the University of Tübingen, studying with Michael Maestlin, an early believer in Copernicus's heliocentric theory. Kepler not only believed in the reality of heliocentricism but also found religious significance in it, defending Copernicanism from both a theoretical and a theological standpoint.

Official mathematician in Graz and Prague

Although he aspired to a pastor's career, in 1594 Kepler accepted a mathematics position in Graz, Austria. As district mathematician, he was expected to prepare an annual calendar and astrological prognostication; his 1595 edition made his reputation as an accomplished astronomer. His classroom use of geometry led to the 1596 publication of *Mysterium cosmographicum* (*Mystery of the cosmos*), the first published defence of Copernicus's system, which described God's structure of the Universe based on the five Platonic solids, as well as the relationship between a planet's speed and its distance from the Sun. Readers of *Mysterium* included the Italian mathematician Galileo Galilei, who admitted to being a secret Copernican, and the Danish nobleman Tycho Brahe, who suggested that data from his observatory might help refine Kepler's theories.

In 1600 Protestants were banished from Graz. To escape the growing religious tension, Kepler travelled to Prague, where Tycho had become court mathematician to Emperor Rudolf II. Following Tycho's death in 1601, Kepler assumed his post. More than a quarter of a century would elapse before he completed Tycho's proposed Rudolphine Tables, using his predecessor's data. To justify his salary in the meantime, Kepler identified projects that culminated in two masterworks: *Astronomiae pars optica* (The optical part of astronomy) of 1604 and *Astronomia nova* (New astronomy) of 1609. The first volume – with its discussion of refraction; the inverse-square law for the intensity of light (then no more than an intuitive guess); reflection by flat and curved mirrors; pinhole cameras; and the anatomy of the eye, human vision and correcting short- and far-sightedness with lenses – founded modern optics. The second set out the principles of planetary motion that would bear and establish his name.

Tycho's data on the orbit of Mars led Kepler to propose in *Astronomia nova* the first two of what are now called Kepler's three laws of planetary motion: that each planet moves not in a circular but in an elliptical orbit, with the Sun at one focus; and that the line connecting the planet and the Sun sweeps out equal areas in equal times. Negotiations with Tycho's heirs, who maintained rights to the data, delayed publication of *Astronomia nova* until 1609. In the event, Tycho's son-in-law Frans Tengnagel wrote a preface warning Kepler's readers to beware the author's physical arguments. Despite this cautionary notice, *Astronomia nova* remains a mathematical masterpiece. Kepler's laws of planetary motion are correct, although Newton's gravity later superseded Kepler's inaccurate claim that a magnetic force from the Sun causes planetary motion.

News soon reached Prague that Galileo had used a new invention – the telescope – to study the heavens. Galileo sent Kepler a copy of his 1610 Sidereus nuncius (Starry messenger), hoping the imperial court mathematician's favourable assessment would bolster his reputation. Lacking a telescope, Kepler was unable to confirm Galileo's observations, but he issued *A Conversation with the Starry Messenger*, noting particularly Galileo's discovery that Jupiter had four moons. Opponents of Copernicanism had argued that if Earth travelled around the Sun, its moon would be lost; Jupiter, however, retained the moons revolving around it. Kepler thus became the first astronomer to endorse Galileo publicly. Kepler's 1611 *Dioptrice* also explained how the combination of convex and concave lenses in the Galilean telescope works, and described what is now called the Keplerian telescope, whose two convex lenses yield higher magnification than Galileo's instrument.

Kepler flourished in Prague, but in 1611 his imperial patron Rudolf II succumbed to madness and was stripped of all effective power by his brother Matthias, who became emperor the following year when Rudolf died. Matthias renewed Kepler's appointment as imperial mathematician, but allowed him to leave to become mathematician to the Estates of Upper Austria in Linz, away from the political and religious difficulties then engulfing Prague.

Success and strife in Linz

Kepler's fourteen years in Linz were as productive as his time at the imperial court had been. The position permitted him to pursue research and continue work on the Rudolphine Tables. Kepler supplemented his income by preparing prognostications and ephemerides, tables providing daily planetary positions. His 1613 book on the six-petalled snowflake includes his work on the densest possible packing of objects – as in the arrangement of oranges at fruit stands – which was proven correct only in the late 20th century. In 1615, Kepler's *A New Stereometry of Wine Casks* became the first book ever published in Linz. Its surprising subject notwithstanding, it contributed to the development of integral calculus later in the century. And in 1618 and 1620, respectively, he published in Linz the first and second volumes of *The Epitome of Copernican Astronomy*, the first textbook presentation of Copernican theory. (The final volume appeared in Frankfurt in 1621.)

However, alongside these professional successes, Kepler experienced a series of personal difficulties in Linz. His first wife, Barbara, had died before he left Prague, and he had to interview eleven possible successors before finally settling in 1613 on a suitable replacement, Susanna, aged twenty-four. Then, in 1615, he learned that his mother, Katharina, had been accused of poisoning another woman and would later face trial, accused of witchcraft. In 1617–18, two of his daughters with Susanna and a stepdaughter died in quick succession. And in 1619 his avowal that not only Lutheranism but also Calvinism and Catholicism contained religious truth, as well as his unorthodox views on the Eucharist, put him in conflict with the Lutheran Church.

Nevertheless, he continued to work on the *Rudolphine Tables* and was able to concentrate on another major effort. *Harmonices mundi libri V* (Five books on the harmony of the world), published in 1619, includes Kepler's third law of planetary motion. It describes how a planet's orbital period relates to its distance from the Sun, that the period is as the three-halves power of the mean distance. This was later the basis of Newton's first demonstration that the force keeping planets in orbit, the force of gravity, varies inversely as the square of a planet's distance from the Sun. It thus accommodates both the orbit of the innermost planet, Mercury, which orbits in 88 days, and of the outermost planet, Neptune, discovered only in 1846, which orbits in 164 years. (Neptune completed the first orbit since its discovery only in 2010!)

The turmoil of war

Kepler's last dozen years played out beneath the shadow of the Thirty Years War (1618–48), which began as a conflict between Catholics and Protestants in the Holy Roman Empire, but broadened into a European struggle for political domination. Soon after the war began, Holy Roman Emperor Matthias died. Months later Archduke Ferdinand II, a Catholic extremist, succeeded Matthias, ordering forced conversion to Catholicism in Bohemia and Austria. Kepler's ongoing work on the *Rudolphine Tables* obligated him to the imperial court, but would he be able to fulfil his commitment to complete the astronomical tables while remaining a Protestant?

Resolution of this crisis was interrupted in 1620, when Kepler left for Württemburg to assist in his mother's defence. His efforts were in vain, for Katharina was found guilty and imprisoned for fourteen months. After her release in 1621, Kepler returned to the relative peace of Linz. His appointment as imperial mathematician was confirmed, and his personal safety ensured, but just as the printing of the *Rudolphine Tables* was about to commence, a Protestant insurrection erupted in the city when Emperor Ferdinand II ordered that all Protestants convert or be expelled. Though exempt from the order, Kepler was caught up in the disorder. In summer 1626, peasants besieged Linz, igniting a fire in its outskirts. The manuscript of the *Rudolphine Tables* was spared, but fire destroyed the printer's press. After the siege was lifted, Kepler requested permission to leave to find a haven where the *Rudolphine Tables* could be printed. That honour fell to Ulm (where Albert Einstein would be born more than two and a half centuries later), about 100 miles from the Bavarian city of Regensburg, where Kepler now established his family.

To enable users to calculate planetary positions thousands of years forward or back, the *Rudolphine Tables* use logarithms, only recently invented. Along with the telescope, the tables made it possible in 1631 to observe a transit of Mercury across the Sun for the first time in history. In 1629 Kepler published a pamphlet calling attention to the upcoming transit (and to a transit of Venus, not visible in Europe). Kepler's prediction of the transit of Mercury ranks as one of the most spectacular predictions in the history of science. (He missed predicting the 1639 transit of Venus, however, a phenomenon later understood with a modification of Kepler's work.)

But Kepler himself would not live to witness a transit of either planet. In December 1627 he presented a copy of the tables to a pleased emperor, whose supreme commander of imperial troops, Albrecht Wallenstein, had suppressed the Protestant revolt. Wallenstein was rewarded with the principality of Sagan, where in 1628 Kepler arrived as Wallenstein's personal mathematician. Within months of Kepler's arrival, the Protestant majority of Sagan was forcibly converted or expelled, though Kepler was spared. At a 1630 meeting in Regensburg of the imperial officials called electors, Wallenstein was stripped of his command. Fearing for his future, Kepler travelled to Regensburg to assess matters. He fell ill en route and died there on 15 November 1630 in a house that is now a Kepler museum. His final publication, begun when he was a student in Tübingen but appearing posthumously in 1634, is a treatment of astronomy as it might be practised by creatures on the Moon. Kepler's *Dream* has been called an early piece of science fiction. A seminal thinker of the Scientific Revolution thus also founded a popular literary genre.

Galileo Galilei

Laying the foundations of modern science

(1564–1642)

Philosophy is written in this grand book – I mean the Universe – which is always open to our gaze. But it cannot be understood unless you first learn the language it's written in. It is written in the language of mathematics.

GALILEO GALILEI, *THE ASSAYER*, 1623

In 1989, NASA launched the *Galileo* spacecraft to study Jupiter and its moons, honouring the man whose telescope first revealed those satellites four centuries earlier. Galileo Galilei also pioneered the study of nature through experimentation, measurement and mathematical calculation. Because this methodology characterizes all scientific research to this day, Einstein dubbed the Italian astronomer 'the father of modern physics – indeed of modern science altogether'.

Born in Pisa, Tuscany, in 1564, Galileo studied at the city's university. He was appointed in 1589 to a mathematics position, in which he undertook experimental research that challenged traditional ideas that had been passed down from the Ancients. Dropping weights from a great height (supposedly the Leaning Tower of Pisa) and letting balls roll down inclined planes, Galileo disproved Aristotle's assertion that heavier objects fall faster than lighter ones, attributing the difference in their rates of fall to the buoyancy of air; in the absence of air, he argued, all objects would fall at the same rate. (In 1971, *Apollo 15* astronaut David Scott demonstrated on the airless Moon that a feather and a hammer fall to the surface together.)

Making waves in the Venetian Republic

Shortly after his appointment in Pisa, in 1592 Galileo moved to the University of Padua in the Venetian Republic, where he taught courses in Euclidean geometry and Aristotelian cosmology, and tutored students in practical mathematics. He invented a geometric and military proportional compass, supplementing his income by training purchasers to calculate with it. During frequent visits to nearby Venice, where he observed tidal patterns, Galileo developed a theory of the tides based on Nicolaus Copernicus's heliocentric model of the Sun at the centre of the Universe. He believed that interactions between Earth's daily rotation and yearly revolution could produce an oscillation of the water in seas sufficient to cause tides. While Newtonian mechanics later showed that this theory was incorrect, its importance lies in Galileo's mechanical explanation for tides, without resorting to a mysterious force periodically attracting the ocean.

In the summer of 1602, Galileo resumed his earlier studies of motion. His experiments with inclined planes and pendulums led to the conclusion that the distance an object falls from a resting position increases as the square of the elapsed time. By 1608, he understood how motion is accelerated during an object's fall – that the increase of speed is proportional to time. He also demonstrated that projectiles follow a parabolic curve, enabling him to compute a cannon's range.

In October 1604, Galileo observed a nova (Latin for 'new'), a newly visible star. To conform to Aristotelian theory that the heavens were perfect and unchanging, Padua's philosophers proposed that the nova was closer than the Moon, not in the heavens. Galileo, however, argued that because the nova did not appear to move in projection against the stars at different elevations above the horizon, it was farther away than the Moon. Anonymously, Galileo critiqued a Florentine philosopher's book that claimed that the nova had always been present in the heavens, but had remained undetected until a lens in the 'crystalline sphere' moving in the sky suddenly made it visible. (Interestingly, Einsteinian general relativity in the late 20th century has explained a phenomenon known as gravitational lensing that makes some extremely distant objects, which would otherwise be unseen, visible.)

In 1609, Galileo heard that the Dutchman Hans Lippershey had invented a device that made distant objects appear close. In August of that year, by using a plano-convex lens as the objective and a plano-concave lens as the eyepiece, Galileo produced a nine-power telescope. He presented a second, improved version to the governors of Venice, who assembled atop the bell tower of San Marco to admire a view of ships more distant than previously visible. They rewarded him with a lifetime appointment to the University of Padua for his achievement. Within months, Galileo had devised a twenty-power telescope and later a thirty-power instrument, with which he was able to observe the Moon, Jupiter's satellites and stars.

The discoveries reported in Galileo's 1610 *Sidereus nuncius* (Starry messenger), including a rough, mountainous lunar surface and four tiny moons circling Jupiter, challenged Aristotelian principles. Galileo's depiction of a rough lunar surface contradicted Aristotle's notion that all heavenly bodies were perfect, and his observation of moons circling Jupiter refuted the claim that everything revolved around Earth. Galileo's telescopic discoveries convinced him that the Copernican system was preferable to the Aristotelian. He thanked God 'for being so kind as to make me alone the first observer of marvels kept hidden in obscurity for all previous centuries'. The astronomical community did not unanimously embrace Galileo's discoveries or inferences, however. Notable among those who did was Johannes Kepler, then mathematician at the court of the Holy Roman Emperor in Prague.

Court mathematician in Florence

Longing to return to his native Tuscany, and preferring research and writing to teaching, Galileo had been courting a patron in Florence for some time. He had dedicated his 1606 book on the military compass to Prince Cosimo de' Medici, who became grand duke of Tuscany in 1609. In *Sidereus nuncius*, Galileo christened Jupiter's four moons the 'Medicean stars' in honour of Cosimo and his three brothers. By June 1610, his flattery had paid off and he was appointed philosopher and mathematician to the Grand Duke of Tuscany, with an annual salary of a thousand scudi, a substantial sum at the time. In September, Galileo left liberal Padua in the Venetian Republic for a position in the Medicean court in Florence, fatally much closer to Rome and its Inquisition. He was also appointed chief mathematician at the University of Pisa, without teaching responsibilities.

Continuing his telescopic studies, Galileo interpreted the change in diameter he perceived in Venus as its distance from Earth varied, and the full set of phases he observed, as proof that Venus moves around the Sun rather than below it, as in Ptolemy's theory, or above it, which was also possible before the discovery of its phases. He also saw two small stars next to Saturn, which later appeared as 'ears' or 'handles', but were not identified as rings until the Dutch astronomer Christiaan Huygens did so in 1659.

In spring 1611, Galileo spent two months in Rome, where Pope Paul V granted him an audience. He met Christoph Clavius, dean of mathematicians at the Collegio Romano (Roman College), the training centre for the Society of Jesus. Although Clavius rejected Galileo's assertion that the Moon's surface was rough, even after confirming his telescopic observations, other Jesuit mathematicians supported Galileo's conclusions. In Rome, Galileo was inducted into the Lincean Academy, one of the world's oldest scientific societies. He displayed its lynx logo on his publications, two of which the Academy sponsored and published. The first, *Letters on Sunspots*, was written in response to the German Jesuit mathematician Christoph Scheiner's observations; sunspots, which Galileo had already detected, were another challenge to Aristotle's ancient and unquestioned assertion that heavenly bodies were perfect. The battle for priority between Scheiner and Galileo made the latter no friends in the Church. The Lincean Academy also later sponsored Galileo's *The Assayer* of 1623, which skewers a Jesuit mathematician at the Roman College for his interpretation of comets; in fact, the Jesuit, who followed Tycho Brahe's 1577 interpretation of comets as bodies existing in the heavens beyond the Moon, was correct, while Galileo's dismissal of comets as optical illusions in vapours rising from the Earth was wrong. Dedicated to the new pope, Urban VIII, who as Cardinal Maffeo Barberini had been a patron of the Academy, *The Assayer* sports the Barberini family crest on its title page. *The Assayer* also contains Galileo's assertion that humans can comprehend the Universe only through God's language, mathematics.

While Galileo was lionized by significant Romans, his reception among Tuscan philosophers upon his return was frigid. After the appearance in 1612 of Galileo's *Bodies in Water*, three University of Pisa philosophers joined forces with his Florentine critics to discredit the man who challenged their Aristotelian world view.

Arraigned by the Inquisition

The 1613 appointment of Galileo's friend and student Benedetto Castelli as professor of mathematics similarly distressed Pisa's philosophers. Warned by a university official to avoid teaching Copernicanism, Castelli was peppered with questions by the grand duke's mother about the reconcilability of Copernican views with biblical assertions. Seeking advice, Castelli wrote to Galileo, who responded that the Bible and the Universe, both creations of God, are equally true; when observations of the natural world contradict the Bible, however, observed truth must prevail over figurative scriptural language.

Castelli distributed copies of Galileo's letter, which found their way to two Dominicans, and thence to the Inquisition in Rome.

In 1615, Cardinal Roberto Bellarmino, a member of the Roman Inquisition, advised Galileo to treat the motions of the Earth as hypothetical. Late that year Galileo travelled to Rome to attempt to prevent a condemnation of the 'Copernican opinion'. His reception was cool, but believing he was in no immediate danger, he began speaking publicly for Copernicanism and his own tidal theory. The matter came to the attention of Pope Paul V, who instructed Cardinal Bellarmino to warn Galileo on threat of imprisonment to abandon Copernicanism and 'in the future not hold, teach or defend it in any way either by speech or writing', to which Galileo consented on 26 February 1616. In March, Copernicus's *De revolutionibus orbium coelestium* (On the revolutions of the heavenly spheres) was placed on the Index of Prohibited Books pending 'correction' according to Church requirements. Galileo left Rome three months later, carrying a certificate signed by Cardinal Bellarmino, affirming that he had not been punished but had been informed that Copernicus's work was 'contrary to Holy Scripture ... [and] cannot be defended or held'. Galileo later understood this certificate to limit him to hypothetical discussions of heliocentrism, without claiming it was real.

When Grand Duke Cosimo II died in 1621, Galileo worried about the loss of his patron. The election of his friend Cardinal Barberini to the papacy in August 1623, however, delighted him. In spring 1624, he met six times in Rome with the new Pope Urban VIII. Urban upheld the ban on Copernicus, but permitted Galileo to compare the Copernican and Ptolemaic systems in writing, provided he refrained from asserting that Earth's motions were real. Believing he had papal support, Galileo undertook to write his masterpiece, casting his exploration of Copernicanism in the form of a dialogue among three fictional figures. Galileo gave the name Salviati to the character representing Copernicus, after a deceased friend in whose Tuscan estate Galileo had recuperated from several illnesses; the character representing men of common sense was named Sagredo; and the one standing for Aristotelian devotees was Simplicio. That name was a double entendre: Simplicius of Cilicia was a commentator on Aristotle who flourished around 530, but the name also connotes simplemindedness, of course.

In 1630 the Church censor declared he would approve the completed manuscript, if Galileo made prescribed revisions. The censor reported the pope's aversion to the inclusion of tides in Galileo's chosen title, *Dialogue on the Tides*. The book that was eventually published in Florence in February 1632 was thus called simply *Dialogo di Galileo Galilei ... sopra i due massimi sistemi del mondo Tolemaico, e Copernicano* (*Dialogue of the two great world systems: Ptolemaic and Copernican*). It minimized evidence from the tides for Earth's motions and focused instead on the pope's claim that human intellect cannot limit God's power.

Despite Galileo's compliance with the censor's demands, the pope soon abandoned him. In August 1632, Urban prohibited further printing of the Dialogue and appointed a special commission to examine the book. It found a number of points objectionable, including Galileo's disobeying the order not to hold, teach or defend the Copernican opinion. The pope had Galileo summoned to Rome to appear before the Holy Office. The Inquisition also ordered his attendance before its inquisitors, which he did in January 1633, following months of illness. Under interrogation, he insisted that he did not support the Copernican opinion and claimed that Bellarmino's certificate permitted him to speak hypothetically about Copernicanism. The inquisitors rejected his argument. In June, the Inquisition convicted Galileo of being 'vehemently suspect of heresy', placed the *Dialogue* on the Index, and banned publication of any work – past or future – by him. The pope also decided to imprison him for an indefinite period.

At first, Galileo was placed under house arrest at the residence of the Tuscan ambassador to Rome. He was then transferred to the Siena home of Archbishop Ascanio Piccolomini. There he was inspired by visitors to begin his last book, *Two New Sciences*. Covering materials science and the science of motion, it maintained the form of a dialogue between Salviati, Sagredo and Simplicio. By summer 1638, when the book appeared in Holland, Galileo had gone blind, although he was able to continue his correspondence with the help of a secretary. He petitioned the Inquisition to be freed, but the request was denied. He was, however, allowed to move to his house in Florence so that he could be closer to his physicians. Less than four years later, on 8 January 1642, Galileo died in his villa in Arcetri, near Florence.

Posthumous recognition

Pope Urban VIII forbade eulogies or monuments to Galileo after his death. A century later, however, Church doctrine had been relaxed. In 1737, Galileo's remains were transferred from a small crypt in the rear of Santa Croce in Florence to a monument in the main part of the church. In the 1820s, the Catholic hierarchy began to permit publication of books presenting the Earth's motion as fact; and in 1835, the Index of Prohibited Books was the first not to include any books by Copernicus, Kepler or Galileo.

Rehabilitation of Galileo continued under the last 20th-century pope. Addressing the Pontifical Academy of Sciences in 1992, Pope John Paul II did not apologize, but said the inquisitors who condemned Galileo, working in good faith with the knowledge available at the time, did not properly distinguish between the Bible and its interpretation. 'This led them unduly to transpose into the realm of the doctrine of the faith, a question which in fact pertained to scientific investigation.' In 1993, however, Cardinal Ratzinger, who in 2005 became Pope Benedict XVI, said that the outcome of Galileo's trial was 'reasonable and just'. In 2008, Rome's prestigious La Sapienza

University cancelled a papal visit following student and faculty protests against Ratzinger's statements fifteen years earlier. Nonetheless, the pope announced in December of that year that he honoured those celebrating the International Year of Astronomy, commemorating the 400th anniversary of Galileo's use of the telescope to study the heavens. Pope Benedict remarked that understanding nature's laws can deepen an understanding of God's works.

Isaac Newton
The laws of motion and gravity
(1642–1727)

Nature, and Nature's laws lay hid in night.
God said, Let Newton be! and all was light.

ALEXANDER POPE, *EPITAPH: INTENDED FOR*
SIR ISAAC NEWTON, 1730

The founder of modern physics, Isaac Newton, experienced a difficult and lonely childhood. His biological father, a yeoman farmer, died three months before he was born on Christmas Day in 1642 at the family home, Woolsthorpe Manor in Lincolnshire. When he was two years old, his mother Hannah moved away to marry and live with a much older local vicar. Consequently, Isaac was brought up at Woolsthorpe by his maternal grandmother and a carer. A list of sins that Newton composed when he was nineteen shows that he was an angry child who 'wished death to many' and at one point had wanted to burn down the house where his mother and stepfather were sleeping.

From about 1655, Newton attended The King's School in nearby Grantham, where he lodged at the house of an apothecary. He thrived in this environment and was extraordinarily creative in making wooden toys, clocks and other mechanical devices – activities that his 18th-century biographer William Stukeley called 'playing philosophically'. But in 1659 his mother pulled him out of school so that he could run the family estate. Isaac, however, had no interest in fulfilling the duties expected of him. Fortunately, his extraordinary academic talent had been recognized by a number of individuals, including his headmaster and his uncle, who had been to Trinity College, Cambridge. Despite considering academic life a waste of time, Hannah allowed him to return to the grammar school to prepare for university.

Youthful brilliance in Cambridge

Arriving at Trinity College in the summer of 1661, Newton experienced a traditional education based on the writings of the philosopher Aristotle. However, in the spring of 1664 he attended the lectures of the first Lucasian Professor of Mathematics, Isaac Barrow, whose reduction of physics to mathematics left a lasting impression on him. He now turned decisively away from the old-fashioned curriculum and immersed himself in the new 'mechanical' philosophy of advanced thinkers such as René Descartes, Nicolaus Copernicus and Johannes Kepler. Over the next two years, he made seminal discoveries in optics, mechanics and mathematics, working mostly at home in Lincolnshire because the plague had descended on Cambridge.

By the end of 1666 he had become the first to describe techniques for calculus (that is, differentiation and integration) by means of the analysis of infinitesimally small entities that he called 'fluxions'. He was also the first to state the binomial theorem of elemental algebra, which allowed the expansion of the form $(a + b)n$ using a formula that worked for all values of n, including negatives and fractions. It was during this period too, when, apparently prompted by seeing a falling apple, he compared the attraction exerted by the Earth at its surface with that required to keep the Moon in orbit. He found that both phenomena 'pretty nearly' obeyed a law by which the force exerted by the Earth on other bodies was inversely proportional to the square of the distance between them, although the result was not sufficiently exact for him to make it more widely known.

At about the same time, Newton discovered by way of a brilliant series of experiments that white light was heterogeneously composed of more basic rays, each of which had its own colour and its own index of refraction. An offshoot of this research was his invention of a working reflecting telescope, which produced images via a highly polished mirror rather than by refraction through a lens.

In 1667, Newton returned to Cambridge and was made a fellow of Trinity College. But his academic success was only beginning. Over the next couple of years, he wrote up and refined his mathematical researches into a paper, 'On analysis by equations infinite in number of terms', and his efforts were soon rewarded when he was elected to the Lucasian chair upon Barrow's resignation in 1669. Two years later, as Newton rewrote his lectures on optics and his work on fluxions for publication, Barrow brought him to the attention of the Royal Society by showing its members a reflecting telescope that Newton had built. He now sent them a paper that conveyed his discovery of the heterogeneity of white light by means of a 'crucial experiment'. Not only did this overturn both modern and ancient belief that white light was modified in its transition from one medium to another, but also Newton boasted that he had now achieved the incredible feat of making the science of colours *mathematical*.

Newton made a firm distinction between those claims that could be proved with absolute mathematical certainty and those that could not, condemning the latter as mere 'hypothesis' or 'conjecture'. His arguments held little sway with Robert Hooke, author of the famous *Micrographia* of 1665 and the dominant presence of the Royal Society. Hooke believed that light was a wave or pulse that travelled through an invisible medium, or 'ether'. He agreed with Newton on the reality of the phenomena he had described, but still believed that the colours were the result of a modification of the white light performed by the prism. He asserted that Newton's theory was merely a 'hypothesis', a claim that seriously upset the Lucasian Professor.

Against his better judgment, in 1675 Newton was persuaded to release a version of his own private beliefs concerning natural philosophy in the form of a 'Hypothesis of Light'. In this fascinating text he gave a detailed account of his understanding of the various cosmological roles played by an ether, which accounted for light, sound, electricity, magnetism and gravity. This work led to a second dispute with Hooke, who told a number of people that Newton had taken most of what he had written from his own *Micrographia*. Always quick to detect a slight, Newton accused Hooke of taking all his work from Descartes and was only becalmed when Hooke told him that his own views had been misrepresented. In a famous reply Newton remarked that Hooke had indeed done some good work, adding: 'If I have seen further it is by standing on the shoulders of giants.'

Alchemy and theology

The disputes that resulted from his early entry into the Republic of Letters made Newton give up his plans to publish his optical and mathematical work, and he increasingly committed himself to other studies, such as alchemy. In one text he argued that metals 'vegetated' or 'grew like trees' into the Earth according to the same 'laws' that governed the development of living things. Both were due to a 'latent spirit' that energized other processes such as fermentation, nourishment and chemical operations. Newton also devoted himself to theology. By the late 1670s he had developed a sophisticated and deeply Protestant view of history. In one project, probably composed in the mid-1680s, he argued that the ancients believed in a Newtonian cosmos, and worshipped around a central Vestal fire in imitation of the Solar System. This practice, he claimed, was proved by the shape of the ruins at Stonehenge and Avebury and was 'the most rational religion of all' before Christianity.

Undoubtedly the major concern throughout his adult life was the unravelling of the mysteries of prophecy. Working within the Protestant apocalyptic tradition in which the pope was the Antichrist and Catholicism was the religion of Satan, Newton's enquiry was informed by an exceptionally radical brand of anti-trinitarianism (that is, he believed that the notion of a Holy Trinity was a deliberate fabrication). According to Newton, Roman Catholics such as the 'crafty politician' Athanasius of Alexandria were abetted by the Devil, who came down to Earth in the 4th century after Christ, and they imposed their unintelligible and corrupt version of Christianity on a gullible world. Newton lived and worked in a society that would have been horrified by such views, and at the very least he would have been a social pariah had they been known to his contemporaries.

Principia Mathematica

At the end of 1679 Hooke wrote to Newton about the elements of celestial dynamics. During this correspondence, Hooke suggested that the movements of planets and their satellites could be determined by 'compounding' inertial movement in a straight line with an attractive force that pulled bodies from this motion. He also proposed that this force was inversely proportional to the square of the distance between the two bodies. As we have seen, Newton was aware of the inverse-square law but does not seem to have realized the significance of Hooke's other points about orbital motion until after he had dealt with a major celestial event that took place in 1680.

At the end of that year the so-called 'Great Comet' appeared, disappearing behind the Sun at the end of November and being followed swiftly by another comet early the following month. The Astronomer Royal John Flamsteed wrote to Newton in January 1681 to say that he had predicted its return and that the two comets were therefore one, the November comet having been turned in front of the Sun by means of magnetic repulsion. Newton, who at this stage believed that they were two separate comets, replied that the known paths of both comets were inconsistent with their being only one *if* it turned in front of the Sun. If it were one comet, then it had curved round behind the Sun, but there was no known physical mechanism for this.

In any case, Newton doubted that the force emanating from the Sun was magnetic, since he believed that heated magnets lost their power.

At a time when it was supposed that natural philosophy required the articulation of physical causes in order to properly explain phenomena, the only plausible alternatives to magnetism were a vague etherial 'fluid' and the notion of a whirling 'vortex' that had been described in the 1630s and 1640s by Descartes. When Newton announced his theory of Universal Gravitation in his masterwork *Philosophiae naturalis principia mathematica* of 1687, known as the *Principia*, he explicitly denied that space contained either an ether or a vortex. Such mechanisms left no room for God (who sustained a cosmos that constituted an absolute frame of reference) to intervene in his own creation – which he had to do on occasion. Having kicked away the crutch of a physical mechanism to explain his notion of Universal Gravitation, Newton incurred the criticism of many contemporary scientists, but ultimately succeeded in changing what it meant to explain natural phenomena.

The immediate stimulus for the composition of the *Principia* had come as a result of a visit from Edmond Halley in 1684. When pressed, Newton claimed that he could show that an elliptical planetary orbit implied an inverse-square law but was unable to provide a demonstration of this until November of that year. Within twelve months he had discovered that all bodies, no matter how small, possessed the power of attraction and attracted other bodies according to the equation $F = G (m_1 m_2 / r^2)$ (where G is a gravitational constant and r is the distance between the masses m_1 and m_2). Alongside universal gravitation, Newton thereby introduced the modern notions of force and mass, along with his three laws of motion.

The final structure of the *Principia* consisted of three books, the first two of which were treatments of various 'virtual mathematical worlds' in which different laws of nature operated. The second book dealt with motion in media such as fluids, and the third, 'De mundi systemate' (On the system of the world), dealt with the laws of nature as they actually existed in our own cosmos. For the first time Newton gave adequate explanations of tides, cometary motion and the shape of the Earth, and accounting for the orbit of the (now single) Great Comet played a central role. Newton's work was quickly recognized as a work of genius. The most gifted natural philosophers and mathematicians tried to master its contents, and its difficulty became legendary. He was revered by many, though there were dissenters. Just as the *Principia* was about to appear, Hooke became upset by the fact that Newton had not given him proper credit for the hints Hooke had given him about orbital dynamics. Newton himself became irate when he heard of Hooke's complaints, removed some references to him in the draft of the work, and condemned him to Halley as a boastful plagiarist and a mere 'bungler' in mathematics. Similarly, upset by the fact that Newton publicly humiliated him and never acknowledged that he had given Newton crucial data about the orbits of planets and comets, Flamsteed came to see Newton as a pathological tyrant who was obsessed with maintaining the flattery of his idolatrous admirers.

Public life and bitter controversies

In 1687 Newton publicly defended Cambridge University against efforts by the Catholic King James II to insert a Catholic priest into Sidney Sussex College. Two years later, in the wake of the Glorious Revolution, Newton became member of the English parliament for Cambridge University. For the next few years he tried unsuccessfully to obtain a public office in London, but he continued to work intensely on a number of different topics. For example, he sought to show in a number of 'classical' scholia that the ancients had known that God was the immediate cause of gravity, but had hidden this and other truths from the vulgar in mysterious and obscure language. Eventually, in 1696, Newton was appointed to the position of warden of the Mint.

He turned what had been a sinecure for his predecessor into a committed effort to track down the so-called clippers and coiners who were treacherously debasing English coin, and his job occasionally obliged him to sign the death warrants of wrongdoers. Promoted to master of the Mint in 1699, Newton played a significant role in seeing through the amalgamation of the Scottish and English mints leading up to the Act of Union that created Great Britain in 1707.

In 1703, he received the ultimate accolade in British science by being elected president of the Royal Society, and he was knighted two years later. Foreigners gained great prestige by promoting his views outside Britain, and by the 1720s the Newtonian system was dominant in British and Dutch universities and cities. It took at least another two decades for his doctrines to become widely accepted in Italy and France.

Initially, however, the notion that there was some sort of mysterious 'attraction' acting between all bodies in the universe seemed incredible and unscientific to great European natural philosophers such as Gottfried Leibniz and Christiaan Huygens, and Newton's theories led to a number of disputes with rivals throughout his life. Leibniz had visited England in 1673 and 1676, and by the second visit had devised a very different version of calculus than Newton's. At this stage, he and the Cambridge mathematician had a good relationship, expressed in two letters written by Newton to Leibniz in 1676. The mutual respect would not last, however. Leibniz published the rules of calculus in 1684, but evidence of Newton's own work in the area would not appear for another two decades. In the meantime, some of Newton's followers began to suggest that Leibniz's calculus was inferior to Newton's, that their hero had been first to devise it, and even that Leibniz had gained key hints about Newton's discovery during his visit to London in 1676. In 1712 and 1713, the issue exploded in a series of bitter exchanges between Newtonian and Leibnizian lieutenants. This bad-tempered dispute was complicated by the fact that Leibniz was the librarian and effectively court philosopher to the Hanoverian regime that was to continue the Protestant succession (in the person of George I) when Queen Anne died in the summer of 1714. To Leibniz, the Newtonian system was crass, not least because of the ludicrous doctrine of attraction, and also because God had to repeatedly and perversely intervene in order to shore up his own creation. Newton thought that Leibniz, like Descartes, had devised a system that was so perfect that there was no need of God. He also equated Leibniz's opaque metaphysical subtleties with the doctrines of those who had corrupted the simple truths of Christianity.

Despite these debates, Newton's theories continued to dominate much of the intellectual landscape. The appearance of his Opticks in 1704 allowed a much richer set of his doctrines to be discussed and promoted. Consisting mainly of old material, he appended to it a set of 'Queries' in which he revealed his private views on the existence of a whole set of 'active principles' that governed phenomena such as growth and our ability to move our own bodies. In later editions he would add further Queries dealing with the phenomena of chemistry, electricity and magnetism, surprisingly invoking the sort of etherial explanations that had been present in his 'Hypothesis' of 1675.

In the last years of his life, Newton performed many of his administrative duties in a perfunctory way, but sustained his intense interest in theology and chronology. By the time of his death in 1727 he had for many decades been a scientific legend, held in the highest regard by the British state and heralded as the founder of Reason. Despite recent revelations about his occasionally despicable behaviour, historians agree that Newton towered intellectually above his contemporaries as no other since, and most concur with Halley that no mortal will ever approach closer to the gods.

Michael Faraday

Seminal experiments in electromagnetism

(1791–1867)

Science itself is not the principal thing – we are
men and ought to have human feelings.

MICHAEL FARADAY, AS RECORDED IN
JOHN TYNDALL'S DIARY, 5 OCTOBER 1853

The scientific enterprise covers a wide variety of activities. These include research in the making of discoveries and constructing new knowledge about the natural world; using scientific knowledge and methods for practical and technological purposes; communicating that knowledge to others, which in turn is related to the ideological role of science in society; the development and implementation of science policy; and the administration of scientific institutions. As a general rule, scientific practitioners tend to concentrate on only one of these aspects. What distinguishes Michael Faraday, and makes him one of the most recognized scientists of all time, is that he covered all of these areas of science in his career with the highest possible quality and became one of the most famous men in Europe. His origins, however, would not have suggested this career or outcome.

A committed Christian

Faraday was born in south London, to which his parents had moved a few years earlier from Westmorland in north-west England. His father was a blacksmith who belonged to the Sandemanians, a very small sect of neo-Calvinist literalist Christians to which Faraday remained fully committed for his entire life. Coming from a relatively poor background and not belonging to the Anglican Church meant that Faraday could not go to university. Instead, from 1805 to 1812 he was apprenticed to a London bookbinder, though during this period he attended scientific lectures and performed some limited chemical experiments. Towards the end of his apprenticeship, he made the extraordinary decision not to pursue the safe occupation of a bookbinder, but instead determined on a scientific career. To this end, he attracted the attention of Sir Humphry Davy, who, following his marriage at the age of thirty-four to a wealthy widow, was about to retire as professor of chemistry at the Royal Institution of Great Britain. Faraday was appointed laboratory assistant in 1813 and spent virtually his entire career at the Royal Institution, rising to be director of the laboratory in 1825 and having the Fullerian Professorship of chemistry created especially for him in 1833.

Faraday's research into electromagnetism during those years, including his discoveries, made in the basement laboratory of the Royal Institution, of electromagnetic rotations (1821) and induction (1831), in effect resulted in the invention of the electric motor, the transformer and the generator. From the late 19th century onwards, this work was seen, rather simplistically, as laying the foundations of electrical engineering and thus of much of the modern world. While such a view is no longer tenable, the celebrations surrounding the centenary of induction (including a two-week Faraday exhibition at the Royal Albert Hall and a commemorative address by the British prime minister) contributed significantly to Faraday's enduring fame – including his appearance on the reverse of the £20 banknote in the 1990s.

But Faraday's most significant contribution to our understanding of the natural world was his formulation of the field theory of electromagnetism. This followed from his discoveries, in 1845, of the magneto-optical effect and diamagnetism – that is, he experimentally demonstrated that magnetism affected the behaviour of light and that all matter was susceptible to magnetic force. From at least the early 1830s Faraday had been strongly opposed to the idea that matter comprised indivisible chemical atoms; and by 1834 he had ceased to believe in the usefulness of the concept of matter altogether on the grounds that all we can study is force – weight, electrical repulsion, etc. Faraday interpreted even some of his earliest experimental work, such as rotations, in terms of lines of force. By the early 1840s, he came to see matter as points in space where lines of force met. This suggested that all matter was structurally similar; but at that time only three substances were known to possess magnetic properties, meaning that magnetism was something of an anomaly. Throughout 1844 and 1845, Faraday concentrated his experimental efforts on remedying this, and his eventual discovery of the magneto-optical effect and diamagnetism demonstrated that magnetism was a universal property of matter, akin to gravitation.

These experimental discoveries gave him the confidence in 1846 to begin his formulation of field theory, which described how the fields of electricity and magnetism interacted with one another. Though initially qualitative, field theory, by solving the pressing engineering problem of building a telegraph cable from Ireland to Newfoundland across the Atlantic Ocean, came, first in Britain, but later elsewhere in Europe, to displace the mathematical theories of electromagnetic action that had been developed by savants such as André-Marie Ampère. In the mathematical hands of William Thomson (later Lord Kelvin) and of James Clerk Maxwell, Faraday's field theory became, and remains, one of the cornerstones of modern theoretical physics. Although not a mathematician, indeed sometimes doubting the value of mathematics in natural philosophy (complaining once to Maxwell about the 'hieroglyphics' that he used), Faraday was nevertheless a theorist of the highest order. Maxwell saw that Faraday's field approach was essentially geometrical and could therefore be subjected to the rigours of Cambridge mathematical algebraic analysis. In the view of Albert Einstein, writing in 1936, 'the electric field theory of Faraday and Maxwell represents probably the most profound transformation which has been experienced by the foundations of physics since Newton's time'.

Faraday's ability to formulate the field theory depended on his experimental discoveries of the magneto-optical effect and of dia-magnetism – the property of an object that causes it to create a magnetic field in opposition to another one that is applied to it from an external source, thus generating a repulsive effect. In turn, these experiments were entirely contingent on another aspect of Faraday's career, that of scientific adviser. One of the original functions of the Royal Institution, founded in 1799, was to provide scientific advice to those who required it, mostly, but not exclusively, the state and its agencies. Faraday continued this agenda providing advice to institutions such as the East India Company, the Admiralty, the Home Office, the National Gallery and, most important of all, Trinity House, the English and Welsh lighthouse authority – after 1836, when he was appointed their scientific adviser, almost a fifth of Faraday's extant correspondence concerned lighthouses.

In the second half of the 1820s, Faraday had worked for a joint Royal Society–Board of Longitude committee charged with improving optical glass for use in telescopes. Faraday was not able to achieve this goal, however, and by 1829 had become so frustrated with the project that he opened negotiations to be appointed professor of chemistry at the Royal Military Academy. However, Davy, who had been behind the project and tended to exploit Faraday's abilities, died in May and so Faraday was soon able to abandon the glass project, which for the following fifteen years he regarded as a complete waste of his time. However, the project did have some collateral benefit: in 1845 he used a piece of lead-borate glass that he had made in the 1820s to discover the magneto-optical effect. In similar vein, the light source he used for this experiment was a very powerful oil lamp that he was in the process of testing for Trinity House. These and other examples illustrate the close connections between Faraday's research and his practical work.

Enlightening the public

The other main function of the Royal Institution was to communicate science to a middle-class and aristocratic audience. Davy had established the Royal Institution's initial reputation for providing highly popular lectures and Faraday inherited this role, proving even more successful. He founded the Friday Evening Discourses, the weekly, hour-long lectures by eminent scientists that became one of the major vehicles for imparting science in the early Victorian period and continue to this day. In these lectures, Faraday demonstrated to the members of the Royal Institution and thus, via print media, to the rest of the world, the major scientific discoveries that he had made in the laboratory. Following his strong opposition to the craze of table-turning, mesmerism and seances in the early 1850s, he coorganized a course of lectures on the value of scientific education and gave evidence supporting this view to the Royal Commission on education. His deep concern with this issue probably led him to allow his last two series (out of nineteen) of Christmas lectures for young people to be published. *The Chemical History of a Candle* must be the most popular science book ever published; since 1861, it has never been out of print in English and has been translated into at least a dozen other languages.

As a consequence of his research, his lectures and his practical work, Faraday became one of the most famous men of the day (and indeed after). He was a personal friend of Prince Albert, the Prince Consort, who arranged for him to be given a grace-and-favour house at Hampton Court, where he spent an increasing amount of time from 1858, dying there in 1867; he was one of the eight foreign associates of the French Academy of Sciences, the supreme accolade of recognition before the establishment of the Nobel Prize; and he was twice offered the presidency of the Royal Society, the leading position in British science. But, unlike Davy who had failed spectacularly in that job, Faraday was not attracted to such celebratory admiration. On both occasions, he refused the position, regarding it as a corrupt and corrupting office, commenting on the second occasion that had he accepted he 'would not answer for the integrity of my intellect for a single year'.

Nevertheless, although Faraday claimed that he remained humble before his Sandemanian God and argued for humility in scientific research, he possessed an ego that he usually kept under strict control. It expressed itself, unconsciously it would appear, in a number of ways, most notably in the sheer number of images of him that exist in all conceivable media: oils, pastels, marble, drawings, prints and above all photographs (he announced the invention of that technology in a lecture in 1839). Faraday, at some level, despite the requirements of his religious beliefs and his approach to studying the world, wanted society to know that he existed. Such tensions caused by this internal conflict may well account for his creativity and obsessiveness in everything that he did, and thus explain why he contributed fundamentally to our understanding of the world.

James Clerk Maxwell

The electromagnetic nature
of light and radiation

(1831–1879)

*The advance of the exact sciences depends upon the
discovery and development of appropriate and exact ideas,
by means of which we may form a mental representation
of the facts, sufficiently general, on the one hand, to stand
for any particular case, and sufficiently exact, on the other,
to warrant the deductions we may draw from them by the
application of mathematical reasoning.*

JAMES CLERK MAXWELL, *FARADAY*, 1876

Three names are key in the history of modern mathematical physics:
Isaac Newton, Albert Einstein and James Clerk Maxwell. The natural
philosophy of Maxwell, who was born in Edinburgh, was influenced by the Scottish Enlightenment, and also by the Industrial
Revolution, German philosophy and romanticism, Cambridge
University mathematical physics, and Victorian imperial culture. It
was also at the intersection of what are now several distinct fields
of enquiry: mathematics, experimental physics, metaphysics, logic,
philosophy of language, rhetoric, cognitive psychology, aesthetics,
ornamental design, natural and personal theology, electrical and
mechanical engineering, political economy, and the physiology of
vision and movement. In physics, Maxwell's many contributions
spanned colour theory and optics, the mechanics of elastic solids and
fluids, astronomy, the molecular physics of gases and, most notably,
electromagnetism. His work is characterized by a combination of

mathematical sophistication, self-conscious use of language and methods, a spirit of unification, and a regulated employment of the imagination at the service of understanding natural phenomena and abstract mathematical theories. Maxwell had a 'marvellous interpenetration of scientific industry, philosophic insight, poetic feeling and imagination, and overflowing humour', according to his friend and biographer Lewis Campbell.

A family of diverse talents

Maxwell was a scion of the remarkable Clerk family of Penicuik in the county of Midlothian, Scotland; the surname Maxwell was added by law to his father's name Clerk when the latter inherited an estate formerly belonging to a family named Maxwell. His great-great-grandfather studied medicine under the Dutch physician Hermann Boerhaave, with whom he also composed music; collaborated with the Scottish mathematician Colin Maclaurin, Newton's expositor; was a discriminating art connoisseur and avid collector; designed one of the great Palladian houses with his protégé and fellow freemason, the Scottish architect William Adam; and wrote innovatively about architecture and mining as underground architecture. His great-grand-uncle was an accomplished draughtsman and etcher, specializing in landscapes and architecture, and an expert in mineralogy, geology and mining, who contributed to James Hutton's geological work, including diagrams and illustrations for Hutton's *Theory of the Earth*. His uncle led the parliamentary commission that issued the imperial standards of weights and measures. His father was a landowner and lawyer interested in technological advances and science, who designed a novel printing press, and took his son to meetings of the Royal Society of Edinburgh and the Royal Scottish Society of Arts. His cousin Jemima Wedderburn, with whose family he lived after his mother's early death in 1839, married the mathematician Hugh Blackburn, the Glasgow colleague of William Thomson (the future Lord Kelvin). An accomplished watercolourist and Victorian illustrator, she was admired by her artist acquaintances Ruskin, Millais and Landseer.

Thus, from an early age, Maxwell's life was part of a rich social, cultural and intellectual network. Religion was important, too. Maxwell experienced both his father's Scottish Presbyterianism and his aunt Jane Cay's Episcopalianism. These emphasized the value of gaining knowledge of the world, so as to know and praise the creator through his works, by encouraging the use of the artificial to explore and understand the natural. In particular, this relied on the design and construction of imaginary and material models and instruments alongside the experimental manipulation of substances and objects.

Between the ages of ten and sixteen, Maxwell attended the prestigious Edinburgh Academy, where he quickly became a skilled draughtsman and poet, geometry expert and amateur scientific experimenter. His love of models in science can often be traced to his childhood games and toys. In 1847, he entered the University of Edinburgh, where he registered to study literature and then studied natural philosophy for three years as the protégé of his relative, the natural philosopher James Forbes, cofounder of the British Association for the Advancement of Science. He learned chemistry, Cambridge-style mathematics, rhetoric and Aristotelian and Kantian logic and metaphysics. He then moved to Cambridge, where he came under the influence of the omniscient scientist William Whewell, master of Trinity College, who was an idealist historian and philosopher of science, architectural historian, poet, moralist and educator. Trinity offered Maxwell a Germanic metaphysical culture of romanticism, classics, theology and intellectual and emotional bonding. He was recruited into the secret debating society known as the Cambridge Apostles, and he also taught in the local branch of the newly opened Working Men's College, inspired by Christian socialists led by theologian F. D. Maurice.

After graduating in 1854, at the same time as his father's death, Maxwell took over the running of his inherited Scottish estate. In 1856, he took a post at Marischal College in Aberdeen, where he continued teaching workers and married the principal's daughter, Katherine Dewar (the marriage was childless). But in 1860 the position was cancelled, and Maxwell moved to King's College, London. There he joined the British Association's efforts, led by Thomson (Kelvin), to establish new British electrical standards. Thomson devoted his scientific life to a widely applied but narrowly conceived mathematical, experimental and technological practice. In his view, the challenge in trying to understand, predict and manage the new Victorian economy was to quantify as much as possible. Maxwell, for his part, wanted to apply the same thinking to natural phenomena, rather than manufacturing – the measurement of which also required a language of uniformity, generality and precision based on conventions. These conventions were crucial to the new telegraphic cable networks that sustained Thomson's Glasgow laboratory, his production of precision measurement instruments, and indeed the British empire. In 1871, Maxwell was appointed professor of experimental physics at Cambridge and director of the new Cavendish Laboratory, which was purpose-built to his own design, incorporating an electro-technical workshop, in part to continue the standardization of electrical units. He was working there at the time of his death in 1879 from stomach cancer, at the same young age as his mother.

In the area of colour theory, as a student Maxwell researched colour-blindness, which was rife in Edinburgh. He established a 'geographical method' for studying the old artistic problem of how colours could be created by mixing primitive colours – typically red, blue and yellow. This led to a general and precise objective representation of a subjective phenomenon: a coordinate system (land map) locating each colour as a point on a colour triangle, and an algebraic equation for each colour in terms of the quantities of three new primitive colours, red, green and violet. Maxwell's theory was based on experimental results. As a Cambridge undergraduate, he had designed a set of spinning colour disks to study the quantitative mixture of colours with standards supplied by David Hay (who also defended the mathematical rationality of relations between colours), and later a colour box like a doll's house to analyse the spectral decomposition of different colours. He introduced the idea of colour field from the notion of field of view in the German physiological literature. He thus vindicated Thomas Young's 1802 proposal of a three-receptor theory for the physiology of colour vision. To illustrate Young's theory at a Royal Institution lecture in 1861, Maxwell contrived the projection of the first colour photograph.

Electromagnetic waves

The combination of mechanics and optics appeared in his most celebrated contribution to science: his mathematical theory of electromagnetism in terms of electric and magnetic fields of forces and energy. Maxwell borrowed Faraday's experimental results and descriptions concerning the relation between electricity and magnetism, the rotational nature of magnetism, and the notion that electric and magnetic forces act contiguously and along curved lines of force in tension between polar opposite states. In order to capture mathematically the physics of contiguous action, as opposed to the Newtonian model of action at a distance, Maxwell (and also Thomson) employed differential equations as representations of contiguous causal action.

Maxwell thus formulated a unified mathematical theory of electromagnetism based on fields of force and energy defined at every point in space, predicating his ideas on the existence of a mechanical ether pervading the universe like an invisible muscle that could store and communicate this energy, which could be measured by its capacity to do work. He sought understanding of electromagnetism and the ether through imaginary mechanical models of stress and energy in the form of tubes of fluid flow. In 1861, he presented a molecular model of the electromagnetic ether with microscopic rotating vortices in rolling contact. The theory predicted the existence of electromagnetic waves and the value of their velocity of propagation, which was very close to the experimental value of the velocity of light. From this equivalence, Maxwell correctly asserted that light must be an electromagnetic wave, and that optics could therefore be reduced to electromagnetism. This single discovery laid the foundations of modern physics, for it led to the development of special relativity and quantum mechanics in the early 20th century. For this reason, Maxwell is often described as the most important scientist of his age – and, after Newton and Einstein, the greatest physicist ever.

Over time, Maxwell replaced his commitment to the incomplete molecular mechanical representation of transmission of electromagnetic action with a safer adherence to general explanations from general principles. His ideas were collected in 1873 in his greatest publication, A Treatise on Electricity and Magnetism. His application of molecular physics extended from astronomy to microscopic molecules. He was able to explain the stability of Saturn's rings in terms of the velocity of an indefinite number of small independent particles orbiting around the planet at different distances. This rotating model strengthened his interest in statistical molecular studies of macroscopic properties such as temperature, pressure and viscosity in his dynamical theory of gases. This work with thermodynamic behaviour in turn inspired his molecular rotating model of the mechanical transmission of contiguous electromagnetic action in the ether. Although eventually all of these new molecular models partially failed – until the arrival of quantum mechanics in the 20th century – they cemented the use of a 'statistical method' based on probability theory, which describes the group properties of populations of identical molecules and the behaviour of large systems; the 'historical method', by contrast, describes the properties and evolution of individual molecules at the microscopic level. Maxwell himself is credited with using the term 'statistical' mechanics to describe this approach to physics for the first time. His so-called imaginary 'demon' (a term coined by Thomson) was a fictional entity of molecular size, his scientific version of Alice, which he created in a thought experiment to show that it was possible to reverse the flow of heat from hot to cold at the molecular level. This established that the irreversible macroscopic processes described by Thomson's second law of thermodynamics can have only a statistical certainty.

Maxwell's contributions to physics thus represent the grand culmination of the tradition of natural philosophy and the mechanical view of the world, in terms of a continuous theory of forces and a discrete theory of matter. They were superseded only in the early 20th century with the emergence of Einstein's theory of special relativity and quantum physics.

Albert Einstein

Thought experiments in space, time and relativity

(1879–1955)

An hour sitting with a pretty girl on a park bench
passes like a minute, but a minute sitting on
a hot stove seems like an hour.

ALBERT EINSTEIN'S EXPLANATION OF RELATIVITY
GIVEN TO HIS SECRETARY TO RELAY TO REPORTERS
AND OTHER NON-SCIENTISTS

Most areas of physics and its associated technology have been fundamentally affected by the work of Albert Einstein. They range from the workings of the Universe and the accuracy of the satellite Global Positioning System to the structure of the atom and the origins of the laser. Einstein himself regarded his quantum theory of 1905, arising from his study of laboratory data on the photoelectric effect – the way in which certain metals emit electrons when light falls on them – to be his 'revolutionary' achievement; and it was indeed this work for which he was awarded a Nobel Prize in 1921. But current quantum theory owes more to the work of other physicists, such as Niels Bohr, Max Born, Werner Heisenberg and Erwin Schrödinger, who developed Einstein's quantum theory in the 1920s in ways that he did not entirely approve.

Einstein's most individual and enduring achievement is undoubtedly the theory of relativity – presented initially in 1905 in its 'special' form, which applies to uniform motion, and subsequently in 1915 in its 'general' form, which incorporates the acceleration due to gravity.

Einstein's theory produced the greatest change in our understanding of space and time since Isaac Newton stated his laws of motion and gravitational attraction in the 17th century. In general relativity, there was no longer any need to posit a mysterious ether filling empty space or the unsatisfactory concept of gravity's 'action at a distance'. The Universe became instead a space–time continuum, in which, so to speak, matter tells space how to curve, while space tells matter how to move.

Einstein created his theory of special relativity while working at a full-time job as a patent clerk in Switzerland. From conception of the idea to completion of his article, 'On the electrodynamics of moving bodies', took a mere five to six weeks in May and June 1905, Einstein told a biographer in 1952. 'But it would hardly be correct to consider this as a birth date, because the arguments and building blocks were being prepared over a period of years, although without bringing about the fundamental decision.'

Inauspicious beginnings

In the Einstein family tree, there was no hint of any intellectual distinction. His father Hermann was an easy-going businessman who was not very successful in electrical engineering, while his mother Pauline, a fine piano player but otherwise not gifted, also came from a business family, which ran a profitable grain company and was wealthy. Although both sides of the family were Jewish, neither was orthodox or devoted to reading the scriptures – an ignorance of Hebrew that Einstein would regret when he became a Zionist in the 1920s.

Nor was there much sign of distinction in Einstein as a child. Albert was born in Ulm in the Kingdom of Württemburg, then part of the German Empire, the first of two children. He was a quiet baby, so quiet that his parents became seriously concerned and consulted a doctor about his not learning to talk. But when his sister Maja was born in 1881, the two-year-old Albert is said to have asked promptly where the wheels of his new toy were? His ambition was apparently to speak in complete sentences: first he would try out a sentence in his head, while moving his lips, only then repeat it aloud. The habit lasted until his seventh year or even later. The family maidservant dubbed him 'stupid'.

However, he displayed an early interest in science. When he was four or five, his father showed him a magnetic compass. Albert saw how the compass needle determinedly pointed in one particular direction, even though no one had touched it. Here was the first of many encounters with the concept of a field – in this case the Earth's magnetic field – which produces seemingly magical action at a distance. The conflict with his unconscious childish assumption that you have to touch an object to make it move provoked wonder and puzzlement. 'Something deeply hidden had to be behind things', he recalled in his autobiography.

Academically he was good, though by no means a prodigy, both at his schools in Germany and Switzerland, and at college in Zurich. Yet Einstein showed hardly any affection for his schooling, and in later life he excoriated the system of formal education in Germany. He disliked games and physical training, and detested anything that smacked of the military discipline typical of the Prussian ethos, as he would make plain to the entire world in abandoning his native country for the United States in the early 1930s with the rise of Nazism, and in his post-war campaigning against nuclear weapons. The main problem with school was probably that Albert was a confirmed autodidact. From a relatively early age, he began reading mathematics and science books simply out of curiosity; at college in Zurich he ranged very widely in his reading, including the latest scientific journals; and as an adult he never read books simply because they were said to be classics, only if they appealed to him. Maybe there is a parallel here with Newton, an eclectic reader who nevertheless does not seem to have read many of the great names of his or earlier times.

Aged twelve, Albert experienced 'a second wonder' – the first being the compass – while working through a small book of Euclidean plane geometry. The 'lucidity and certainty' of the geometrical proofs, based on ten simple axioms, set Einstein thinking on the true relationship between mathematical forms and the same forms found in the physical world. The very word geometry, he noted, was from the Greek for 'earth-measuring', which implied that mathematics 'owes its existence to the need which was felt of learning something about the behaviour of real objects'.

The breakthrough of relativity

Some time in 1895–96, Einstein, aged sixteen, began to think originally about moving bodies, space and time. From Newton's laws and James Clerk Maxwell's equations of electromagnetism, he would ascend via special relativity in 1905 to the heights of the field equations of general relativity in 1915. He achieved this not by overturning Newton or Maxwell but rather by subsuming them into a more comprehensive theory, somewhat as the map of a continent subsumes a map of an individual country.

It was his deeply held youthful view that throughout the physical world the laws of mechanics, and indeed the laws of science as a whole, must be the same – 'invariant' in scientific language – for all observers, whether the observers are 'at rest' or are moving uniformly. Near the beginning of _Relativity: The Special and the General Theory_, Einstein's 1916 introduction to the subject for the general reader, he describes a simple but profound observation. Imagine standing at the window of a railway carriage that is travelling uniformly – at constant velocity, not accelerating or decelerating – and letting a stone fall onto the embankment, without throwing it. If air resistance is disregarded, you, though you are moving, see the stone descend in a straight line. But a stationary pedestrian, that is someone 'at rest', who sees your action ('misdeed', says Einstein) from the footpath, sees the stone fall in a parabolic curve. Which of the observed paths, the straight line or the parabola, is true 'in reality', asks Einstein? The answer is both paths. 'Reality' here depends on which frame of reference – which system of coordinates in geometrical terms – the observer is attached to: the train's or the embankment's. Moreover, he says, there is no such thing as the absolute frame of reference for the Universe, against which all velocities may be measured, as found in 'classical' physics. For Newton, this reference frame was God; for Maxwell, it was the ether. For Einstein, the absolute frame of reference did not exist.

But if this first postulate about the invariance of the laws of nature was physically correct, it must apply not only to moving bodies but also to electricity, magnetism and light: the electromagnetic wave of Maxwell, which by 1905 was known from experiments to move at a constant velocity in a vacuum of about 186,000 miles per second, supposedly relative to the stationary ether. This posed a severe problem. While Einstein was content enough to relinquish the ether, which had never satisfied him as a concept, the constancy of the speed of light was another matter.

Pushing the barrier of light

He had long reflected on the puzzle of what would happen if one chased a beam of light and caught up with it. In 1905 he concluded: 'If I pursue a beam of light with the velocity c (velocity of light in a vacuum), I should observe such a beam of light as a spatially oscillatory electromagnetic field at rest. However, there seems to be no such thing, whether on the basis of experience or according to Maxwell's equations.' To catch up with light would be as impossible as trying to see a chase scene in a movie in freeze-frame: light exists only when it moves, the chase exists only when the film's frames move through the projector. Were we to travel faster than light, Einstein imagined a situation in which we should be able to run away from a light signal and catch up with previously sent ones. The most recently sent light signal would be detected first by our eyes, then we would see progressively older signals. 'We should catch them in a reverse order to that in which they were sent, and the train of happenings on our earth would appear like a film shown backwards, beginning with a happy ending.' The idea of catching or overtaking light was clearly absurd.

Einstein therefore formulated a radical second postulate: the speed of light is always the same in all coordinate systems, independent of how the emitting source or the detector moves. However fast a hypothetical vehicle might travel in chasing a beam of light, it could never catch it: relative to the vehicle, the beam would always appear to travel away at the speed of light.

This could be true, Einstein eventually realized, only if time, as well as space, was relative and not absolute. In order to make his first postulate about invariance compatible with his second about the constancy of the speed of light, two 'unjustifiable hypotheses' from Newtonian 'classical' mechanics had therefore to be abandoned. The first to go was that 'the time-interval (time) between two events is independent of the condition of motion of the body of reference'. The second was that 'the space-interval (distance) between two points of a rigid body is independent of the condition of motion of the body of reference'. Thus the time of the person chasing the light wave and the time of the wave itself are not the same. Time flows for the person at a rate different from that of the wave. The faster the person's vehicle goes, the slower his time flows, and therefore the less distance he covers (since distance travelled equals speed multiplied by duration of travel). In Stephen Hawking's words, relativity 'required abandoning the idea that there is a universal quantity called time that all clocks would measure. Instead, everyone would have his or her own personal time.' For space, too, there is a difference between the person and the light wave. The faster the person goes, the more his space contracts, and therefore the less distance he covers. Depending on how close the speed of the person's vehicle is to the speed of light, his time dilates and his space contracts in proportion to the time and space of an external observer ('Mission Control'), according to Einstein's equations of relativity. However, like the person who drops a stone from a uniformly moving train and sees it fall in a straight line, not a curve, the person who is chasing the light wave does not himself perceive his watch slowing or his body shrinking; only the external observer sees these effects. To the moving person, everything in the vehicle seems normal. This is because his brain and body are all equally affected by his speed. His brain thinks and ages more slowly, and his retina is squashed in the same ratio as the vehicle; hence his brain perceives no change in the size of the vehicle or his body.

A shock to the Newtonian system

These ideas seem extremely alien to us on first encounter because we never travel at velocities of even a tiny fraction of the speed of light, so we never observe any 'relativistic' slowing of time or con-traction of space. Human movements seem to be governed entirely by Newton's laws (in which the speed of light is a quantity that does not even appear). Einstein himself had to struggle hard in 1905 to accept these relativistic concepts so remote from everyday experience.

With space contraction, he at least had the knowledge of com-parable 1890s proposals by the Dutch physicist Hendrik Lorentz and his Irish counterpart George FitzGerald, although they had a different theoretical basis from his own and relied on the existence of the ether, a concept that Einstein had of course rejected. But the abandonment of absolute time required a still greater leap of the imagination. Henri Poincaré had questioned the concept of simultaneity in 1902 in his *Science and Hypothesis* (read by Einstein on publication). Poincaré wrote: 'Not only do we have no direct experience of the equality of two times, but we do not even have one of the simultaneity of two events occurring in different places.' Indeed Poincaré seems to have come very close to a theory of relativity just before Einstein, but apparently drew back because its implications were too disturbing to the Newtonian foundations of physics. Simultaneity is a very persistent illusion for us on Earth because we so easily neglect the time of propagation of light; we think of it as 'instantaneous' relative to other familiar phenomena like sound. 'We are accustomed on this account', wrote Einstein, 'to fail to differentiate between "simultaneously seen" and "simultaneously happening"; and, as a result, the difference between time and local time is blurred.' Being a generation younger than Poincaré, and having no scientific reputation to lose in 1905, Einstein could afford to be radical in his thinking about time.

Special relativity – including its famous equation $E = mc2$, linking energy, mass and the speed of light squared – was accepted by most physicists by the time Einstein moved to the United States in 1933 to take up a position at the Institute for Advanced Study in Princeton, and it was applied in the calculations for the making of the atomic bomb in 1945. By contrast, general relativity took much longer after 1915 to be fully accepted; and the obsessive quest of Einstein's final decades, from about 1925 until his death in Princeton in 1955, in which he attempted to find a unified theory of gravitation and electromagnetism, is generally regarded as a wasted effort. Today Einstein's theory of relativity, after surviving all manner of ever more precise experimental tests, in space as well as on Earth, is part of the foundations of physics, along with Newton's and Maxwell's laws. After a century's 'continuous flight from wonder' – his own striking description of his creative process – Einstein's esoteric thought world has come to seem almost mundane.

Edwin Powell Hubble

Astronomer of an expanding universe

(1889–1953)

Equipped with his five senses, man explores the universe around him and calls the adventure Science.

EDWIN POWELL HUBBLE, *THE NATURE OF SCIENCE*, 1954

Edwin Hubble was one of the greatest astronomers of modern times. He showed that our Milky Way, of which the Sun is a member, is nothing more than a normal galaxy, and that the objects then known as 'spiral nebulae' are separate galaxies in their own right. By doing so, he revolutionized our understanding of the nature and scale of the Universe.

Hubble had a somewhat varied career to begin with. He was born at Wheaton, Illinois, where he concentrated on mathematics and astronomy, graduating as a bachelor of science in 1910. He then spent three years at Oxford University as one of the first of their Rhodes Scholars; some of his acquired English mannerisms remained with him throughout his life. On his return to the United States, he qualified in law, taught in a high school in Indiana, and gained his doctorate in astronomy at the University of Chicago. During World War I, he served in the US army, rising quickly to the rank of major. He never actually saw active service, but always liked being referred to as 'Major Hubble'. Following the end of the war, in 1919 he was invited to join the staff of the Mount Wilson Observatory near Pasadena, California, where he remained until his death more than thirty years later. His marriage was happy; his wife, Grace, outlived him.

These were exciting times in astronomy. The great 100-inch Hooker reflecting telescope had recently been set up at Mount Wilson. It was not only the largest and most powerful telescope in the world, but was in a class of its own. Hubble was able to make good use of it. It had long been known that the objects referred to as nebulae were of two distinct types: some, such as M.42 in Orion, were obvious gas-clouds, while others, such as M.31 in Andromeda, seemed to be made up of stars. (The M stands for Messier, a French astronomer who in 1781 drew up a catalogue of over a hundred nebulous objects; the M numbers are still used today.) Hubble was confident that the gaseous nebulae belonged to the Milky Way system, but about the starry nebulae he was not so sure. Was it possible that they were entirely separate, and immensely remote? Certainly they were so far away that their distances were too great to be measured by the techniques available at the time. Many of them, including M.31, were spirals resembling Catherine wheels. They had one other notable characteristic: measures made elsewhere, notably by Vesto Slipher at the Lowell Observatory in Arizona, had shown that most of them were moving away from us at high speeds. Slipher had used spectroscopic methods to make his measurements; the light from the starry nebulae was slightly reddened as a result, and this – the well-known Doppler effect – indicated a velocity of recession.

A cosmic controversy

Before he had been at Mount Wilson for long, Hubble had become convinced that the spirals really were independent systems, but other well-known astronomers disagreed. One of these was Harlow Shapley, director of the Harvard College Observatory, who had been the first to measure the size of our galaxy. Another was Adriaan van Maanen, a Dutch astronomer who had been at Mount Wilson since 1912. Van Maanen attempted to measure internal motions in the spirals; he found the stars in them were moving relative to each other, and this meant they could not be as remote as Hubble believed, because even at 100,000 light years, the shifts of the individual stars would have been too slight to be measurable. Therefore, said van Maanen, the spirals were contained in our own galaxy. (This professional disagreement between colleagues at the same observatory was not helped by the fact that they disliked each other intensely!)

Hubble decided to try an entirely different method, using convenient stars known as Cepheid variables. Most stars, including our Sun, shine more or less steadily for year after year, century after century, but others do not; they brighten and fade, some regularly and some unpredictably. The Cepheid variables, named after the best-known member of the class, Delta Cephei, have periods ranging from a few days to several weeks, and they are absolutely regular, so that we always know how they are going to behave; Delta Cephei itself, easily visible with the naked eye in the northern hemisphere of the sky, has a period of 5.4 days – that is, one maximum brightness will be followed by another 5.4 days later. It was also known that the real luminosity of a Cepheid is linked with its period; the longer the period, the more luminous the star. Thus another northern Cepheid, Eta Aquilae, has a period of 7.2 days, and is more luminous than Delta Cephei. It follows that once we measure the period of a Cepheid, we can find its luminosity and therefore its distance – and all Cepheids are very powerful, so that they can be seen across many light years of the galaxy. (A light year, remember, is the distance travelled by a ray of light in one year; not very far short of 6,000,000 miles. Recent measures give the distance of Delta Cephei as 982 light years.)

What Hubble set out to do was to find Cepheids in the starry nebulae, including the spirals. Only the Mount Wilson 100-inch telescope was equal to the task, and Hubble had full access to it. Before long, he was successful. He found Cepheids in several of the spirals, including M.31, and showed that they were so distant that they could not possibly belong to the Milky Way. They were indeed separate galaxies, and his discovery, announced on 1 January 1925, changed our entire view of the Universe. Van Maanen had made a completely honest mistake: when measuring his star plates, he had not taken account of certain photographic effects that gave the appearance of movements that were not real.

Hubble's reputation was made, and for the rest of his life he continued his researches, making other important discoveries. His main assistant was Milton Humason, who began his career as a mule-driver transporting materials up the mountain when the Mount Wilson Observatory was being built and ended it as a world-famous astronomer at the same institution. In particular, they found a link between a galaxy's distance and its speed of recession; the rule was 'the further, the faster'. They discovered that the whole Universe is expanding, though it is not quite true to say that all the galaxies are receding from each other – they form groups, and every group is moving away from every other group. It so happens that our galaxy and the Andromeda Spiral are members of what is called the Local Group, and eventually they will collide, but fortunately for us this will not happen for at least a thousand million years in the future.

Today there is strong evidence that our expanding universe emerged from a 'Big Bang' about 13.8 billion years ago – and that some galaxies had already formed when it was only 10 per cent of its present age. Optical telescopes are now powerful enough to detect these immensely remote galaxies. Astronomers can therefore not only determine the present expansion rate of the Universe – the so-called 'Hubble constant' – but also, by looking at galaxies whose light set out in the distant past, measure how that rate has changed over cosmic history. During the earlier eras, the expansion gradually slowed down: this is the expected consequence of the gravitational pull that galaxies exert on each other. But to the surprise of most astronomers, the expansion was found (in the late 1990s) to have speeded up during the last five billion years. The cause of this –

a force latent in empty space that overwhelms gravity, sometimes called 'dark energy' – remains a deep mystery. How strange now to recall that when Hubble began his studies of Cepheid variables, it was generally believed that the whole of the Universe was contained in the Milky Way.

In addition to his technical papers, Hubble found time to write popular books, of which the best remembered is *The Realm of the Nebulae*. It is fair to say that he was not always the most popular astronomer at Mount Wilson, and his colleagues tended to regard him as rather aloof, but this was certainly not my experience of him; I met him many times after the end of World War II, and to a young English amateur, interested mainly in the Moon, he was unfailingly courteous and helpful. He received almost every honour that the scientific world could bestow, and had he not died suddenly in 1953 he would almost certainly have been awarded that year's Nobel Prize in Physics. He will never be forgotten, and it was surely appropriate that the first major space telescope should have been given his name when it was launched into orbit by the space shuttle Discovery in 1990.

EARTH

In 1968, the astronauts of *Apollo 8* took the first-ever photographs of the whole Earth from space. Risen above the barren, cratered surface of the Moon, the inhabited planet was seen as a sunlit globe floating in blackness with its dark-brown continents and deep-blue oceans wreathed in swirling white clouds. These marvellous images helped to alter the direction of science. Today, few scientists – whether they be mineralogists, seismologists, oceanographers, meteorologists, biologists, or whatever their specialism – question the wisdom of studying Earth as a single, hugely complex system, rather than clinging to science's earlier approach of treating rocks and minerals, mountains and seafloors, icecaps and deserts, rivers and oceans, the atmosphere, fossils and the biota as largely separate areas of study. Without an Earth-system perspective, an understanding of the effects upon these realms produced by human agricultural, industrial and urban activity – the aim of global climate-change science – would be unattainable. One influential thinker, the British chemist and inventor James Lovelock, has even given a name to the Earth system, Gaia, after the ancient Greek Earth goddess. According to the Gaia hypothesis, the Earth is actually alive, a super-organism of which humankind is a part: a most important part, but not the boss.

The pioneers of earth sciences, the Scottish geologists James Hutton and Charles Lyell, the German polymath Alexander von Humboldt and his compatriot meteorologist Alfred Wegener, among others, shared at least some of this modern scientific perception. Each in his travels and expeditions discovered varied geological and other evidence for regarding Earth as a system. Hutton, in the late 18th century, rejected the theistic account of Earth's recent and rapid creation – that is, the belief that God created the Universe and constantly intervenes in its processes – and the catastrophe in the biblical story of the Flood, but substituted a deistic concept of nature – in which God is the creator but does not subsequently act or intervene. The natural world remained the work of a deity, though now a remote and impersonal one, and was designed so as to sustain intelligent human life; and it was eternal, with 'no vestige of a beginning, no prospect of an end'. For Hutton, the Earth was a stable system in dynamic equilibrium, closely analogous to the Solar System of planets circling the Sun.

For later generations, however, as a result of posthumous championing by his fellow Scot John Playfair, Hutton became known principally for his emphasis on the Earth's indefinitely long timescale and hence the slow operation of geological processes such as sedimentation, erosion, uplift and volcanism. Lyell built upon these ideas, and concluded that such processes acted uniformly throughout Earth's history, neither faster nor slower than today, without feeling the need to postulate past catastrophic change. Thus, the present was the key to the past, and the Earth was a self-balancing non-progressive system. Lyell's evidence for this gradualism, in his 1830–33 *Principles of Geology*, exerted a powerful influence on Charles Darwin when he travelled the world on board the *Beagle* between 1831 and 1836, although it was soon challenged by Louis Agassiz's discovery of a catastrophic ice age.

Darwin's other major influence was Humboldt's celebrated account of his expedition to South America between 1799 and 1804. Although Humboldt made no earth-shattering discoveries in his long life, his immensely wide-ranging observations and researches, and his eloquent discussion of their interrelatedness – the 'knowledge of the chain of connection, by which all natural forces are linked together, and made mutually dependent upon each other' – make him the founding father of what later became known as the science of ecology.

Wegener took a similarly interrelated view. His profoundly controversial theory of continental drift, published in 1915, was initially based on the remarkable resemblance of geological formations and fossils on opposite sides of the Atlantic, in South America and Africa. But he lacked a physical mechanism for the drift. However, over the four editions of his *Origins of Continents and Oceans*, Wegener gathered further multi-disciplinary evidence of the past linking of continents. It is appropriate that the clinching evidence, which appeared only in the 1960s long after Wegener's death, came from a barely explored realm: the seafloors of the Atlantic and Pacific, where continental movement was directly observable at rift valleys. This quickly gave rise to the most important development in Earth-system science, the theory of plate tectonics.

James Hutton
The Earth's stable system
(1726–1797)

For having seen a succession of worlds, we may from this
conclude that there is a system in nature; in like manner as,
from seeing revolutions of the planets, it is concluded, that
there is a system by which they are intended to continue
those revolutions.

JAMES HUTTON, *THEORY OF THE EARTH*, 1795

There is irony and paradox in the claim, still commonly made by geologists though no longer by historians, that James Hutton was the uniquely influential 'father' or founder of geology, or that his theory of the Earth is the foundation of modern science. For his legacy to later generations, which was profound, was quite different from what he himself believed was most important about the Earth.

Hutton was the son of a merchant living in Edinburgh. As a young man he worked towards a medical career, first at the university in Edinburgh and then in Paris, the centre of the scientific world, and Leiden, where he acquired a medical degree. Significantly, in view of his later work, his dissertation was on the circulation of the blood, a natural steady-state system. Back in Scotland, with a growing interest in chemistry, he and a friend started a successful factory to produce ammonium chloride for industrial use. Later he acquired farms near Edinburgh, and was much involved in the movement to improve agricultural practices. These activities gave him an income that enabled him to settle as a bachelor in Edinburgh, enjoying a convivial life with men of similar intellectual interests.

Hutton was in many ways a typical philosopher of the Enlightenment, a friend of David Hume, Adam Smith and other famous Edinburgh intellectuals. His most ambitious published work was his massive 1794 *Principles of Knowledge*, on the fundamentals of epistemology. His 1795 *Theory of the Earth*, for which he was later (and still is) most often remembered, was intended to illustrate his wide-ranging philosophy; with the same intention, he also published on topics in physics and chemistry, and at his death he left in manuscript a major work on the *Principles of Agriculture*. His concept of the natural world, which he held in common with many other Enlightenment thinkers, was that it is intelligently designed and fit for purpose: designed by a deity, though a remote and impersonal one; and fit for the purpose, above all, of sustaining the life of intelligent human beings, who alone can investigate and appreciate it. This 'deistic' concept of nature was just as religious in character as the culturally dominant Christian theism to which it was tacitly opposed.

A groundbreaking theory

Hutton's deism underlay his belief that the Earth is designed as a 'system' in which all the components interact to maintain a dynamic equilibrium, closely analogous to that of the 'solar system' of planets circling endlessly around the Sun. In another analogy, he thought of the Earth as a 'machine' like a steam engine. Only this kind of steady-state system could ensure the Earth's continuing habitability by human beings. But it did not seem to be in a steady state. Hutton's experience as the owner of farms near Edinburgh led him to believe that the soil – on which, through plant and animal life, human life ultimately depends – is a wasting asset, constantly being washed away. It is only maintained by the continuous disintegration of the underlying rocks, but that material too is eventually washed out to sea. So in the long run the continents are being worn down, and no dry land would be left if they were not somehow being renewed.

Hutton resolved this apparent flaw in the Earth's designful 'system' by conjecturing that the materials eroded from the land and deposited on the sea floor are there consolidated into new rocks, which are later elevated to form new continents. This would complete a vast cycle of processes, and ensure that dry land is always available to support human life.

When in 1785 Hutton first proposed this theory in public, he backed it up with evidence already well known to others with similar interests, though he interpreted it differently. He argued that rocks were consolidated, and new continents uplifted, by the immense power of the Earth's subterranean heat, which was the driving force behind this endless cycle of change. He found signs of this almost on his doorstep, in rocks that he interpreted as products of this intense heat. He then travelled around Scotland, looking for further evidence of the huge power of the Earth's internal heat to elevate new continents out of the ocean floor. Above all, he found evidence for the repeated sequence of processes required by his concept of the Earth in dynamic equilibrium: a 'succession of worlds', of which our present one was just the most recent but not the last. It was, he concluded, a system with 'no vestige of a beginning, no prospect of an end'. This theory, that the Earth is in a steady state, with its natural processes forever operating at unchanging intensities, was later called 'uniformitarianism'.

Hutton's ingenious theory had several consequences that his contemporaries found hard to accept. The least problematic was the long timescale that he took for granted. The Earth's geography seemed not to have changed significantly over the course of recorded human history (that is, since classical times), so Hutton inferred that the processes of erosion and uplift must operate imperceptibly slowly. That the Earth's timescale might be inconceivably vast was widely accepted at this time among educated people, including those who counted themselves religious. But Hutton was rightly taken to be advocating a deistic 'system' that was operating from and to eternity, and that would lose its whole point if human life were not equally eternal. His theory was therefore incompatible with the theistic concept of the created status of the Universe and everything in it, and of the historical character of the Earth and human life. On a more mundane level, Hutton's theory of the consolidation of rocks by fusion under intense heat, rather than by percolating aqueous fluids, seemed highly implausible to chemists, except in the case of certain rocks (now termed 'igneous'), which did indeed appear to have originated as hot melts.

Hutton's *Theory of the Earth* might have been forgotten, along with other eighteenth-century works with similar titles, had it not been championed after his death by his younger Edinburgh contemporary, the mathematician and astronomer John Playfair. But Playfair's 1802 *Illustrations of the Huttonian Theory of the Earth* radically altered the character of Hutton's theory by downplaying its deistic emphasis on designfulness and the primacy of human life, and highlighting instead its assumptions of an indefinitely long timescale and of the correspondingly slow operation of processes such as erosion and uplift. The latter were the 'Huttonian' principles that a later generation of scientists, notably Charles Lyell and (through him) Charles Darwin, was to adopt and put to new explanatory uses.

Charles Lyell

Earth's present as the key to its past

(1797–1875)

As we explore this magnificent field of [geological] inquiry,
the sentiment of a great historian of our times may
continually be present to our minds, that 'he who calls
what has vanished back again into being, enjoys a bliss
like that of creating'.

CHARLES LYELL, *PRINCIPLES OF GEOLOGY*, 1830

Charles Lyell is perhaps best known today as the geologist whose books were read by Charles Darwin during his famous voyage on the *Beagle*, and who thereby influenced the work that led to the younger man's evolutionary theory. But Lyell was far more important in his own right as a leading figure in the development of geology. His greatest work, the *Principles of Geology*, was first published in three volumes between 1830 and 1833, but he continued to revise it in successive editions up to the year of his death. It was, in effect, his side of an ongoing dialogue with other geologists, most of whom strongly criticized some aspects of his work while appreciating and adopting others. The modern earth sciences are, in some important respects, the product of this fruitful interaction.

Lyell came from a family of Scottish landed gentry, but he was brought up in southern England and was, throughout his adult life, a Londoner; he returned only for holidays to the family home on the edge of the Highlands. After his student years at Oxford, he trained in London as a lawyer; he practised briefly as a barrister, before finding that he could earn an adequate income as an author. He married

Mary Horner, a highly educated daughter of Leonard Horner, the first head of University College London; they had no children. His younger friend Darwin, after visiting them at home and seeing her ignored while Lyell relentlessly talked geology, wrote to his fiancée ironically, 'I want *practice* in ill-treating the female sex.' But she was invaluable to him as, in effect, his research assistant, both at home and on their extensive travels on the Continent and, later, in the United States. Lyell, a liberal Whig in British politics, was culturally a cosmopolitan; he was fluent in French, the then international language of science and culture, and felt at home among intellectuals of all nationalities. He was brought up in the Anglican faith, but later in life he regularly attended a Unitarian church in London.

Reconstructing earth's hidden history

Lyell integrated two contrasting intellectual traditions, each of which had already helped to shape the sciences of the Earth. The first he imbibed as a student at Oxford, from lectures by the charismatic professor of geology William Buckland. Buckland in turn had adopted the approach advocated by the great Parisian zoologist Georges Cuvier, who urged geologists to 'burst the limits of time': not to extend the Earth's timescale – which, geologists already agreed, was vast beyond human imagination – but to learn how to reconstruct its long *history* in reliable detail. This required geologists to use fossils and other traces of the deep past as clues, just as historians used documents and artefacts to reconstruct human history. Lyell, like his mentor Buckland, adopted Cuvier's policy: geologists had to become *historians* of the Earth, reconstructing its history from whatever traces of its past survived into the present.

Around the same time, however, Lyell was deeply impressed by James Hutton's model for geology, which saw the Earth primarily as a body governed by the unchanging and therefore *ahistorical* laws of nature. Hutton had earlier portrayed the Earth as a physical system in a steady state of dynamic equilibrium, analogous to the Solar System of planets orbiting the Sun, and this model had been made more accessible to Lyell's generation by the mathematician and astronomer John Playfair. Playfair focused Lyell's attention on what he later called 'modern causes'. These were geological processes – such as volcanoes and earthquakes, erosion and sedimentation – that could be directly observed in action at the present day and then be used to interpret the traces of their former action in the unobservable pre-human past. Other geologists already agreed that 'the present is the key to the past' (as it was put much later), but Lyell became convinced that modern causes were adequate to account for *all* the traces of the deep past, not just some of them. There was no need, he claimed, to infer that there had ever been catastrophic events of a magnitude unparalleled in human experience.

One such alleged event was at the centre of geological debate. Buckland and most other geologists saw widespread physical traces of what they called a 'geological deluge', generally envisaged as some kind of mega-tsunami that had swept over much of Europe, if not worldwide, perhaps around the dawn of human history, but certainly very late in the Earth's much longer history. Buckland identified this with the early human record of Noah's Flood, and used it to support the historicity of the Bible and hence to bolster the acceptability of the new science of geology in conservative circles, particularly at Oxford. Lyell, who was highly critical of the cultural dominance of the Church, opposed this by arguing that the evidence for an alleged deluge, and for similarly catastrophic events even earlier in the Earth's history, could all be explained in terms of ordinary processes.

Readings of the fossil record

Lyell's confidence in the explanatory power of these modern causes was greatly strengthened during his geological Grand Tour of France and Italy in 1828–29. He saw with his own eyes the magnitude of the effects of present-day volcanoes such as Vesuvius and Etna. His fieldwork also gave him evidence that the Earth's crust had been just as dynamic during human history as it had been earlier in the planet's own history. And in Sicily he found a chain of evidence that linked human history to the geological history, giving him a newly vivid sense of the sheer magnitude of the Earth's timescale. Other geologists (including the many who counted themselves religious) already claimed to accept this, but Lyell believed they were failing to appreciate it in practice. He concluded that an adequate sense of the power of modern causes, when combined with an adequate sense of the magnitude of the time in which they had acted, made it unnecessary to invoke exceptional catastrophes of any kind. He argued that all the evidence pointed towards the Earth being in a steady state of dynamic equilibrium, with endless cycles of slow and stately change, and in the long run no overall direction.

Lyell's massive *Principles of Geology*, which he wrote on his return to England, was designed to reinterpret Buckland's kind of geology in these 'Huttonian' terms, using all the persuasive eloquence he had cultivated in his brief first career as a barrister. In Britain, the reading public was struck mainly by the contrast between Lyell's reconstruction of a vastly extended history of the Earth and the constricted timescale traditionally extracted from the biblical texts; in the rest of Europe, the public was better informed about scholarly methods of biblical interpretation, the Earth's vast timescale had long been taken for granted, and Lyell's approach seemed less novel.

Lyell devoted much of the *Principles* to what he called 'the alphabet and grammar of geology'. He used his own observations, enriched with a vast range of published sources and first-hand reports by others, to demonstrate the sheer power of modern causes – given adequate time – to produce even the greatest effects, such as the elevation of new mountain ranges. All this was warmly endorsed by other geologists, who took Lyell's emphasis on the power of modern causes to heart, and increasingly adopted it in their own work. However, they questioned whether *all* the traces of the past could be explained in this way: a few events, such as the geological deluge, might have been genuinely catastrophic in scale, due perhaps to ordinary processes acting at exceptional intensity. Their caution was soon vindicated, for in the 1840s the putative deluge was reinterpreted as a geologically recent ice age. Buckland was among the first to adopt this new idea, whereas Lyell was extremely reluctant to concede the reality of any such catastrophic event.

In the culminating volume of the *Principles*, Lyell used the 'alphabet and grammar' of geology to decipher the traces of the Earth's past history, and to reconstruct its most recent periods as being in all essentials much like the present world. And he interpreted the oldest rocks as having been so altered by what he called 'metamorphic' processes in the depths of the Earth that they could give no evidence for any beginning to the planet's history. So he concluded that the Earth was indeed a system in dynamic equilibrium, as Hutton had argued long before, with no overall direction to its endless round of slow and stately change. Unlike Lyell's effective use of modern causes, his Huttonian model for the Earth was strongly criticized by other geologists: it was this, not his use of modern causes, that was later dubbed 'uniformitarian', though he himself remained the sole example of such a geologist. Above all, other geologists pointed to the increasingly clear evidence, in the fossil record, for an overall direction to the history of life on Earth: most strikingly, among vertebrate animals, fish appeared first, then reptiles, then mammals, and finally humans. Lyell was of course well aware of this, but he had to explain it away, unconvincingly, as the result of a systematically imperfect fossil record. And when he linked the cyclic repetition of the Earth's physical states to a similar kind of repetitiveness in the history of life, other geologists felt his theorizing had gone beyond the bounds of credibility.

In the long run, however, Lyell did help to create a fruitful synthesis between the strongly historical model for the science of geology and the strongly physical model, as exemplified respectively by Buckland and Hutton. The Earth turned out to have a *history* as contingent and as unexpected (even in retrospect) as human history; yet at the same time, all its events could be attributed to *ahistorical* geological processes grounded in the unchanging laws of nature, albeit operating at highly variable intensities. This synthesis was what Lyell's disciple Darwin, who first made his name as a geologist, later transposed so effectively into biology. But Lyell's achievement deserves to be recognized in its own right, and his methods of reasoning about the Earth remain at the heart of modern geology.

Alexander von Humboldt
Adventurous explorer and pioneering ecologist
(1769–1859)

*Nature is a free domain, and the profound conceptions
and enjoyments she awakens within us can only be vividly
delineated by thought clothed in exalted forms of speech....
In considering the study of physical phenomena ... we find
its noblest and most important result to be a knowledge
of the chain of connection, by which all natural forces are
linked together, and made mutually dependent upon each
other; and it is the perception of these relations that exalts
our views and ennobles our enjoyments.*

ALEXANDER VON HUMBOLDT, *COSMOS*, VOL. I, 1845

Although he is little known today, by any measure the German
Alexander von Humboldt was one of the most important scientists
of the modern era. It was Humboldt whose scientific expedition to
the Americas, from 1799 to 1804, inspired a young Charles Darwin
to see for himself the tropics that Humboldt had made famous; and
Humboldt, too, who led Darwin to questions about the origin and
distribution of species, and to the tools by which such questions
could be answered. Throughout the 19th century, Humboldt's name
was a household word, a touchstone for the heroism and humanity of science. He was famous as the romantic adventurer who had
rafted the Orinoco, climbed Ecuador's Chimborazo (thought to be
the highest mountain in the world), escaped jaguars and handled
electric eels. Ralph Waldo Emerson said of him that 'a whole French
Academy traveled in his shoes', and indeed, Humboldt contributed to

scientific fields from botany, zoology and geology, to physiology and geophysics, to geography, anthropology and political economy. He catalysed international science, and the global network of weather stations he founded sparked the study of climate change. In his best-selling books, Humboldt used his adventures to teach radically new concepts in science, and to condemn slavery and imperialism. All human beings were one single species, he wrote, and 'All are in like degree designed for freedom.' Yet his name was never linked to one single transcendent discovery. Newton had gravity; Darwin, evolution; Einstein, relativity. And Humboldt?

One answer could be ecology. The word itself was not coined until after his death, yet Humboldt founded the field of plant ecology, and his writings emphasized the interrelatedness of all elements of nature, including human beings, in the modern, global world. Born in Berlin in 1769, the age of Voltaire, Humboldt died in 1859 even as Darwin was drafting *On the Origin of Species*. He grew up on the family estate at Tegel with his brother Wilhelm (who would himself become a famous philosopher and linguist), both members of the exciting intellectual circle around Kant, Goethe and Schiller. During these years, scientific explorers were returning with stories of exotic peoples, paintings of strange and beautiful landscapes, shiploads of plant and animal specimens, fossils, minerals and curious artefacts of all kinds, unknown to European science. The young Humboldt was fired with the desire to explore the world. His ferocious appetite for knowledge took him through a string of universities – Frankfurt, Göttingen, Hamburg, the Freiberg School of Mines – to study botany, history, political science, geology, chemistry, physiology and languages. Once he inherited his share of the family fortune, he was free to pursue his dream of devoting himself to science; soon after he was on his way to South America with his companion, the botanist Aimé Bonpland.

A boundless search for a world view

What science needed, thought Humboldt, was not more things, bigger collections, but a new way to understand how all things were *interrelated*. Why does the world look the way it does, with continents shaped just so amid oceans charged with global currents circulating between polar ice caps, populated with carbon-dioxide-breathing plants that produce oxygen, and animals that do precisely the reverse? Is electricity the secret of life? (Those electric eels helped him learn the answer.) Why are rocks and minerals everywhere the same, but plants and animals and peoples everywhere different – but different by just a little? Across the globe, human beings grow potatoes and corn, peaches and cherries, wheat and olives and vineyards – yet such plants nowhere grow wild. Then where had they come from? Where had *humans* come from? And why do we vary so much, in appearance and customs and languages? Are all languages related? Are all *humans* related? Are we shaped, perhaps, by our environment – and do we shape it in turn? Why had Columbus sailed west? What was left of the great American civilizations that Europeans had destroyed? How had the gold of the Incas and Mexican silver changed the world economy? What causes tropical diseases, and could they be cured? Would Venezuela and Mexico revolt and form independent republics like the United States?

Such questions and hundreds more in the same vein would form the warp and weft of Humboldt's dozens of books, from scientific monographs to popular essays, which collectively redrew our conception of the planet and sparked dozens of scientific specializations. Though he never taught at a university, he mentored many young scientists, explorers and artists. Yet as one of these beneficiaries, Louis Agassiz, observed, 'Every school-boy is familiar with his methods now, but he does not know that Humboldt is his teacher.' The original mind is hidden from us 'by the very abundance and productiveness it has caused'.

The science closest to Humboldt's heart – what he called 'physics of the Earth', now often called 'Earth systems science' – is today growing rapidly as it integrates ecological and earth sciences, atmospheric and climate sciences, geology and ocean science into a holistic understanding of how our planet works and how human activities are affecting it. One can see the beginnings of this field in two of Humboldt's most important works, the first from the start of his career, the second from the end. After five years exploring the American tropics, Humboldt settled in Paris, then the centre of world science, and set about publishing his findings.

The first volume (there would be thirty in all) was his 1807 *Essay on the Geography of Plants*, a sketch – he called it a 'tableau' – synthesizing his most important observations. In the accompanying illustration, Chimborazo rises in the centre, flanked by dense columns of data. Up the mountainside marches its characteristic vegetation, belt by belt: tropical at the equator, pines and oaks in the temperate middle, alpine plants above, rocks and polar ice at the peak. Each belt represented a climate zone, showing how mountain altitude correlates with global latitude. In each zone, plants combine into typical 'communities', interlinked by such physical characteristics as animal life, elevation, rainfall, light intensity, temperature, soil type and humidity. Each wild plant community has its own distinctive appearance, or 'physiognomy', while humans migrating across the planet have everywhere spread their domesticated plants and animals. The landscapes created by humans and nature thus acting together become part of each human culture, each individual person's mind and heart. The viewer who studies Humboldt's vast picture becomes a participant in it, working actively to interrelate its many elements just like the many scientists whose collaboration made the picture possible. Thus through science and imagination, the reader can, even 'without leaving his home, appropriate everything that the intrepid naturalist has discovered in the heavens and the oceans, in the subterranean grottos, or on the highest icy peaks', live in the past as well as the present, learn to understand nature's great laws, and communicate 'with all the peoples of the earth'.

Humboldt moved to Berlin in 1827, where he spent his last years writing the best-selling *Cosmos*, published between 1845 and 1862, the sum of everything he had learned. The first volume takes the reader on a journey to the stars, passing interstellar wonders on its way to the greatest wonder of all, Earth, a jewel in deep space throbbing with life. The second volume journeys through human history to show how our very concept of 'nature' developed through the millennia in art and literature, science, technology and exploration. Later volumes were to have outlined all that was known of stars and planets, of the Earth and its living systems, but even Humboldt could not keep up with the growth of science, and he died with his great vision unfinished. But his vision could have no end: science is still adding to it today. Hence if one asks what was Humboldt's single great discovery, the answer would have to be the Cosmos itself. What Humboldt shows is the totality of the physical universe, including the way we, human beings, by seeing in that universe its beauty and order, bring it to life as the Cosmos. To Humboldt's view, all nature and all history had cooperated to bring the Cosmos into being, and sustaining it through science, art and poetry would be, he hoped, the future of all humanity.

Alfred Wegener
Meteorologist and proponent of continental drift
(1880–1930)

*If we are to believe Wegener's hypothesis we must
forget everything which has been learned in the
last seventy years and start all over again.*

R. T. CHAMBERLIN, PROFESSOR OF GEOLOGY
AT THE UNIVERSITY OF CHICAGO, 1928

Alfred Wegener became widely known in his lifetime for two seemingly disparate fields of achievement: Arctic climate research and continental drift theory. To him, however, these two endeavours were logical pursuits of his avid interest in the study of climate. Wegener was born in Berlin on 1 November 1880. His education first focused on astronomy, and later on meteorology and climatology. In 1905 he became a meteorological observer at the Aeronautic Observatory near Berlin; and in 1909 he took a position at the University of Marburg, teaching astronomy and atmospheric physics. In 1924 Wegener was offered a professorship at the University of Graz, Austria, a position he had long wanted. By this time, he had already attracted considerable attention – and widespread scepticism – for his theory about how the Earth's landmasses had been formed.

His research in Greenland began in 1906 when he was invited to be the climatologist/glaciologist on a Danish expedition led by Ludvig Mylius-Erichsen. In 1912 he returned with another Danish venture, during which he made one of the earliest crossings of Greenland from east to west by dog sledge. These expeditions established Wegener's reputation in Europe, especially in Denmark and Germany.

While planning the Greenland expeditions, Wegener sought the advice of a leading climatologist, Wladimer Köppen, who was then at work on a classification of world climates. In 1913 he married Köppen's daughter, Else. At first the more experienced Köppen was a mentor to Wegener, but soon the two men began to collaborate on a study of palaeoclimates based on geological and palaeontological evidence of past climates, such as coal deposits, salt deposits, plant and animal fossils, and glaciation.

Wegener observed that a number of identical fossils existed on opposite sides of oceans. He found the same situation for geological formations that start in Africa and continue in South America. Geologists knew of these features, and accepted an explanation that plants and animals had migrated along now non-existent land bridges between continents. But Wegener was developing an alternative theory. He saw that the continents almost fit together like pieces of a jigsaw puzzle, and noted that the geological features joined into a continuous pattern. From now on he pursued the idea of continental drift, tirelessly searching for more geological and palaeontological information to test his hypothesis. In 1915, he published The Origin of Continents and Oceans. In this book, he proposed that all the continents had once been joined as one large landmass, which he called 'Urkontinent' or 'Pangaea', the Greek for 'All-lands'. Gradually Pangaea had split apart and the continents had drifted to their present locations.

A storm of controversy

Wegener's book created a great storm of controversy among earth scientists over defects in the theory. After it was translated into English, the controversy expanded to a global scale, resulting in two international conferences, one in London in 1923 and another in New York in 1926, for the purpose of discussing his theory of continental drift. Wegener attended neither of the conferences.

Geologists in the London conference found troublesome geological shortcomings in Wegener's supporting evidence. They cited areas in which too little of the geology had been mapped to support his proposed connections between continents. All critics rejected the suggestion that tides and the Earth's rotation could provide the force to move continents. The most outspoken critic was the British geologist Philip Lake, who said that Wegener was 'blind to every argument' against continental drift. Harold Jeffreys, prominent mathematician and geophysicist, declared that continental drift was 'out of the question' because no force existed strong enough to move continents. The North American conference was especially bitter because the concept of uniformitarianism – the philosophy formed in the late 18th century by James Hutton and others that the natural laws and processes that presently exist in nature have always existed and operate without changing – was so strong in the thinking of American geologists. This resulted in their conclusion that continental drift was not an ongoing process, so must be rejected. Few recognized the discrepancy in their own logic concerning land bridges, or the lack of evidence for bridges. In short, there was ample need for a new paradigm.

The main missing element preventing support for Wegener's theory was knowledge of a force adequate to push continents over the mantle. A few geophysicists of the time had begun to consider the notion that the mantle behaved rigidly in the short term, but over time could produce a convectional flow sufficient to move the continents. In his fourth edition, Wegener acknowledged that currents in the mantle may play a role in continental movement.

Because most prominent geologists had flatly rejected Wegener's theory, other geologists freely criticized it as well. Continental drift became a subject of jokes suggesting that half of a fossil could be found in America and the other half in Europe. Most geologists viewed Wegener as a crackpot. However, a few geologists and biologists liked the concept of moving continents, because it could answer numerous unanswered questions; but at that time the evidence for movement was not credible.

Though disappointed by their rejection, Wegener's resolute attitude towards critics was to assume they had looked at only part of the picture. He wrote, 'Scientists still do not appear to understand sufficiently that all earth sciences must contribute evidence toward unveiling the state of our planet in earlier times, and that the truth of the matter can only be reached by combining all this evidence.' He never stopped gathering more multi-disciplinary findings and proof to answer criticisms for the next edition of his book. He was certain most critics would be convinced when they saw all the evidence. By the fourth edition, Wegener's book was greatly expanded.

After the American conference, the idea of continental drift lay dormant for forty years until the mid-1960s, more than thirty years after Wegener's death, when new instrumentation and massive government funding, incited by the Cold War, led to global mapping of Earth's sea beds. These new efforts produced abundant evidence of sea-floor spreading and ultimately to the concept of plate tectonics driven by convective currents in the mantle.

During the last two years of Wegener's life his attention returned to his interest in the climate of Greenland. The German Committee for Research Funding, which was designed to provide funds to worthy researchers in the terrible economic time of the 1920s, offered Wegener support for a modest project to study the climate of Greenland. This was an important topic of the day because of the influence of Greenland on Europe's climate, and especially because of its location along possible air routes between Europe and North America.

Wegener saw an unexpected opportunity. He submitted a proposal that greatly expanded the committee's idea into a project involving a year of climate observations at three Greenland stations plus glacial

PPARET igitur ex certissimis observationibus, quod via Planetæ in aura ætheria non sit circulus, sed figuræ ovalis, & quod libretur in diametro parvi circelli, hoc modo; Si post æquales arcus eccentri, Planeta pro distantiis circumferentialibus, $\gamma\alpha, \delta\alpha, \varepsilon\alpha, \zeta\alpha$, hoc est, $\gamma\alpha, \iota\alpha, \lambda\alpha, \zeta\alpha$, quibus circui perfectio innititur, distantias diametrales, $\gamma\alpha, \kappa\alpha, \mu\alpha, \zeta\alpha$,

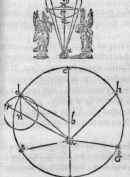

conficiat; ubi ad oculum patet, de semicirculi eccentrici perfectione rescindi tantæ latitudinis lunulam, quanta est quolibet loco differentia distantiarum diversarum, puta $\iota\kappa, \lambda\mu$. Hoc jam obtento, non rationibus a priori, sed observationibus, uti jam dixi; jam speculationes Physicæ procedent rectius quam hactenus. *** Etenim libratio hæc sese accomodat ad spaciũ in eccentrico confectum ; non quidem rationabili seu mentali aliquo modo, ut mens Planetæ æquales arcus eccentrici imperfecti C D. DE. EF. adnumeret æqualibus partibus librationis $\gamma\kappa, \kappa\mu, \mu\xi$. sunt enim hæ inæquales; sed modo naturali, qui nititur non æqualitate angulorum D B C, E B D, F B E, sed ** fortitudine anguli D B C, E B C, F B C, perpetuo crescentis.

Osten. Quid distantia circumferentialis, quid diametralis:

*** Librationis hujus principium probatur esse naturale.

** Quæ sit genuina & ἀναλόγον mensura librationis hujus: sive causa, cur sinus versus anomaliæ eccentri metiatur hanc librationem.

quæ fortitudo fere sequitur sinum Geometris dictum: ubi ascensus continua imminutione sensim in descensum mutatur, probabilius, quam si subito Planeta proram convertere diceretur; quod quidem diximus cap. xxxix etiam experimentis observationum repugnare clarissime. Cum igitur mensura librationis hujus, digitum admodum naturalem intendat: causa quoque naturalis erit; nempe non mens Planetæ, sed naturalis, aut forte, corporalis aliqua facultas.

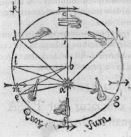

Ac cum sit nobis cap: XXXIX. ex optimis rationibus in præsuppositis, non posse Planetam transitionem facere de loco in locum , nuda contentione virium insitarum , nisi adjuventur aut informentur illæ a vi extranea; cogitandum igitur, si quo pacto, ipsi etiam virtuti Solari, transscribamus hanc librationem ex parte. Id molientes, ad remos nostros jam supra cap: XXXIX. introductos relegabimur . Sit enim flumen aliquod circulare C D E. F G H. in eo sit

Exempla naturalia librationum hujusmodi

In Remis.

Z 3   nauta

1 A page from Johannes Kepler's *Astronomia nova* (New astronomy), published in 1609. The book contains the earliest mention of elliptical planetary orbits around the Sun, and a statement of Kepler's first and second laws of planetary motion.

2 Isaac Newton painted by Sir Godfrey Kneller in 1702, three years before he was knighted by Queen Anne. He was only the second scientist to be given the honour, after Sir Francis Bacon in 1603.

3 Michael Faraday working in his basement laboratory at the Royal Institution, London, some time around 1850. The painting is by Harriet Jane Moore, who documented Faraday's life in watercolours and drawings.

4 Edwin Hubble's discoveries established not only his own fame, but also the international reputation of the Mount Wilson Observatory, California. Here he poses (back left) behind Albert Einstein and other scientists at the site in 1931.

5 Earthrise seen from above the Moon's surface, photographed by the American crew of *Apollo 8* on 24 December 1968. This image inspired scientists to see the world as a single, hugely complex whole, in which a host of scientific fields are intricately connected.

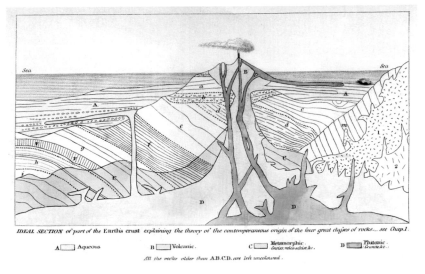

6 Charles Lyell's 'ideal section of part of the Earth's crust', the frontispiece of *Elements of Geology* (1838). He claimed that all the main kinds of rock are still being formed, just as in the remote past, because the Earth is in a steady state of dynamic equilibrium.

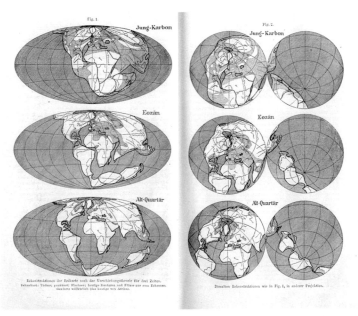

7 Alfred Wegener's *The Origin of Continents and Oceans* (1915) shows the dispersal of continents from a single land mass, Pangaea, through successive time periods. The maps provide equatorial views along with both north and south polar views of the movements.

8 Portrait of Dmitri Mendeleev, creator of the Periodic Table of the chemical elements, painted in 1878 by Ivan Kramskoi, a realist painter who assembled a gallery of portraits of important Russian writers, scientists, artists and public figures.

9 In 1945, Dorothy Hodgkin solved the structure of the penicillin molecule, which she proudly displays in this undated photograph of her and her Oxford laboratory colleagues.

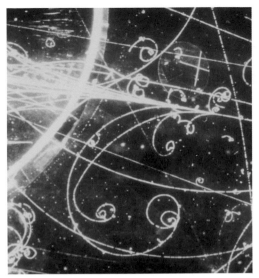

10 Atomic particle tracks in a bubble chamber, a vessel filled with a superheated transparent liquid (often liquid hydrogen), used to detect electrically charged particles moving through it. Nearly half a century after Ernest Rutherford explained the scattering of alpha particles in 1911, the bubble-chamber technique allowed his successors to build on his achievements.

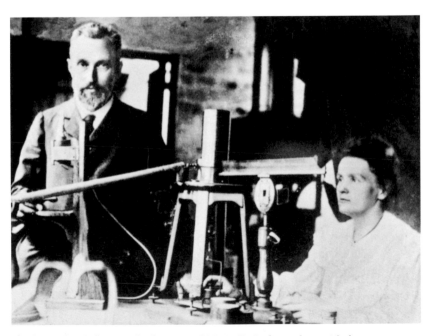

11 This undated photograph of Pierre and Marie Curie shows them with the piezoelectric quartz balance and electrometer they used for measuring radioactivity. Designed by Pierre, this sensitive instrument was operated by Marie, and led to their discovery of radium in 1898.

12 Niels Bohr and Albert Einstein debating in 1925. The two physicists retained the greatest admiration for each other, and were in total agreement on political matters. However, for all their efforts, there was to be no meeting of minds on quantum theory.

13 A posed photograph of Enrico Fermi lecturing at the University of Chicago. On the board, as a joke, he has inverted the value of alpha, the fine-structure constant, which characterizes the strength of electromagnetic force and is one of the supposed fundamental constants of nature.

ice investigations. The ice research was to include measurements of the rate of accumulation of ice and the thickness of ice using seismic reflection techniques for the first time on a glacier. The proposal he submitted requested DM500,000 (£45,000 or US$125,000 at 1929 exchange rates), a very large amount at that time. It is a credit to the high esteem held for Wegener that the Committee quickly funded his request.

Death in the Arctic

In the summer of 1929 he took his principal assistants on a reconnaissance voyage to select an appropriate location on the west coast of Greenland for transporting equipment and supplies up the glacier front. In the spring of 1930 he returned with a ship carrying the entire expedition, including twenty men recruited from Germany, a number of Icelandic ponies, two propeller-driven sledges, and tons of food and equipment.

The late break-up of the ice in the Kamarujuk Fjord caused a six-week delay in unloading. They worked through the summer trying to make up for lost time, and to establish a manned station in the centre of Greenland, called *Eismitte* (Mid-ice). Bad weather and equipment failure prevented them from getting the central station fully provisioned. Two men dug a multi-room ice cave at *Eismitte* and settled in to await the rest of the provisions. Certain essential equipment such as a prefabricated wooden hut, and a shortwave radio never reached *Eismitte*, and it was seriously short of food and fuel.

Winter storms began in September and Wegener decided to make a desperate push to take several dog sledges of food and equipment the 250 miles to *Eismitte*. Because of delays for storms, and disputes with the Inuits helping them, the trip that normally took fourteen days took forty. Most of the Inuits refused to continue the full distance and returned to the west coast. One day after reaching *Eismitte*, Wegener and the one remaining Inuit, Rasmus Villumsen, began the return trip to the west coast. It was the first day of November 1930, Wegener's fiftieth birthday. They never reached the coast. The next summer a search party found Wegener's body in a shallow grave marked with his skis. He was presumed to have died of heart failure due to extreme exertion. Villumsen was never found.

In his lifetime Wegener was best known for his work in the Arctic, but today he is remembered for his remarkable insight that the continents are moving. Because many of the details were wrong, Wegener's critics failed to consider the whole idea of continental drift. Today, however, the concept of continental mobility, transformed into plate tectonics, is fully accepted.

MOLECULES AND MATTER

The idea that matter fundamentally consisted of invisible, indivisible atoms dates originally from ancient Greece. The atomic theory was first proposed in the 5th century BC by Leucippus and his pupil Democritus. However, it was eclipsed for two millennia by Aristotle's theory that all things are formed from only four elements – earth, water, air and fire – and by subsequent alchemical theories, and was revived only in the 17th century in Galileo's corpuscular thinking about all matter being composed of minute particles ('corpuscles') in motion and the mechanical philosophy of René Descartes and Pierre Gassendi.

Not until the late 18th century, through the birth of modern chemistry under Antoine Lavoisier and a little later John Dalton, was there progress in discovering how atoms might aggregate to form matter. Indeed, the common use of the word molecule – the diminutive of the Latin word *moles*, meaning 'mass' or 'lump' – dates from this period. During the 19th century, the existence of atoms and molecules was accepted by the vast majority of chemists and physicists; but it was not considered proven until the 20th century, when scientists could observe atoms using X-ray crystallography, developed by William and Lawrence Bragg, and see the structure of molecules with a form of spectroscopy discovered by Chandrasekhar Venkata Raman.

Though generally regarded as a pioneer of chemistry in the second half of the 17th century, Robert Boyle, a contemporary of Isaac Newton and Robert Hooke (who assisted Boyle with his experiments on air), was as much a physicist. His best-known discovery, Boyle's Law relating the volume and pressure of a gas, is treated as physics.

A dedicated experimenter, committed to the mechanical philosophy, Boyle was nevertheless sympathetic to the idea that corpuscles might have chemical, as well as purely mechanical, properties. Moreover, Boyle by no means rejected the claims of alchemy, as expressed in his book *The Sceptical Chymist*; both he, and Newton, conducted extensive alchemical research and investigations.

A century later, Lavoisier had no truck with alchemy, which was banished from science after about 1800. Lavoisier emphasized a quantitative approach to chemical combination and decomposition with constant use of the chemical balance, founded on the law of conservation of mass: that 'we must always suppose an exact equality or equation between the principles of the body examined and those of the products of its analysis'. By rigorous laboratory-based methods, he established the true nature of combustion and developed a new, rational, chemical nomenclature. In doing so, he discredited the theory first proposed in 1667 by the German alchemist Johann Joachim Becher, which postulated the existence of a firelike substance supposedly present in all combustible bodies – later termed phlogiston. Joseph Priestley had in 1774 discovered what he called 'dephlogisticated air', which combined with substances during burning; Lavoisier, without recourse to phlogiston, identified Priestley's gas as a pure chemical element in atmospheric air, which he named oxygen.

Oxygen was among several important known elements, including carbon, whose relative atomic weights were estimated by Dalton on the basis of their weights as analysed in simple chemical compounds. Water appeared to be about one-eighth hydrogen and seven-eighths oxygen by weight, which led Dalton to assign an atomic weight of 1 to hydrogen and 7 to oxygen, by assuming water's molecular formula to be HO. Although these measured proportions were slightly inaccurate, and the molecular formula in this case was erroneous, Dalton's atomic theory was generally sound. Relative atomic weight was the ordering principle used in the mid-19th century by Dmitri Mendeleev to construct his Periodic Table of the chemical elements, which first revealed the existence of groups of elements with similar properties. Certain gaps in the groups deliberately left by Mendeleev were soon filled by the discoveries of gallium, scandium and germanium – thereby confirming the value of the Periodic Table.

Around the same time as Mendeleev, August Kekulé was the first to understand how organic molecules were built up from chains and rings of carbon atoms – apparently as the result of a dream in which he visualized the ring structure of benzene with six carbon and six hydrogen atoms. His discovery was a crucial breakthrough in the study of carbon chemistry, also known as organic chemistry – the basis of all living things – that led to the development of the modern petrochemical and pharmaceutical industries. By the mid-20th century, complex organic chemical structures comprising hundreds of atoms, for example insulin, had been analysed by devoted X-ray crystallographers such as Dorothy Hodgkin.

Robert Boyle

Experimental investigations
into the nature of matter

(1627–1691)

He made Conscience of great exactnes in Experiments.

AUTOBIOGRAPHICAL NOTES BY ROBERT BOYLE,
AS DICTATED TO GILBERT BURNET, AROUND 1680

Experiment was central to Robert Boyle's adult life. He first discovered the potential of scientific research in the years around 1650 when he lived in the manor house at Stalbridge, Dorset, which he inherited from his father, the Great Earl of Cork, and decided to devote his life to science. Prior to this, he had enjoyed a privileged upbringing, during which he had travelled in Europe and had penned moralistic works. Experiment remained equally predominant in Boyle's later years from 1668 to his death in 1691, when he lived with one of his eminent siblings, his sister Lady Ranelagh, in a house in London's Pall Mall: the latter was equipped with a laboratory in which Boyle carried out experimental trials on an almost daily basis.

Perhaps most important of all, however, were the experiments that he carried out between 1655 and 1668, when he lived in Oxford. It was at this time that he executed the famous series of experiments that formed the subject of his first scientific book, published in 1660: *New Experiments Physico-Mechanical, Touching the Spring of the Air, and its Effects*. Using an air-pump or vacuum chamber, a crucial piece of apparatus designed by his then assistant Robert Hooke, Boyle broke new ground in this book by establishing the characteristics and functions of air, including its role in respiration; in a sequel published in 1662, he adumbrated the law that bears his name – that the relationship between the volume of air and the pressure it is under is a constant. In subsequent books, published right up to the year of his death, Boyle presented equally seminal experimental investigations of the nature of colours, cold and innumerable other phenomena. Perhaps above all, Boyle engaged in chemical experiments, building on and exploiting the procedures for analysing and manipulating substances that were used by practical chemists of his day.

More than anyone, Boyle was a pioneer in devising and implementing controlled experiments, the findings of which he carefully recorded; he also reflected deeply on the nature of experimental investigation, in effect devising a philosophy of experiment. His achievement in this respect is perhaps his chief claim to scientific significance, and the example he provided was all the more important for being championed by the newly founded Royal Society, of which he was one of the earliest fellows.

Corpuscularianism and its complexities

Beyond the actual phenomena that they brought to light, Boyle's experiments served a further purpose, which was to vindicate the mechanical philosophy, the view that everything in the world could be explained in terms of the interaction of matter and motion. In his own words: 'Of the principles of things Corporeal, none can be more *few* without being insufficient, or more *primary* than *Matter* and *Motion*.' Boyle did not originate this doctrine: its pioneers were thinkers like Pierre Gassendi and René Descartes. But he did more than anyone to transform it from an ingenious hypothesis to a doctrine with some plausible empirical basis, not least by adducing experimental evidence: here, the crucial experiment was his reintegration of nitre from its component parts, as expounded in his 1661 *Certain Physiological Essays*.

Equally important was Boyle's more programmatic work of 1666–67, *The Origin of Forms and Qualities*, in which he demonstrated that the explanatory principles of the matter theory inherited from Aristotle which had prevailed for centuries were at worst meaningless and at best superfluous; in fact, all of the phenomena involved could be better explained according to the mechanical or (to use his own preferred term) 'corpuscular' hypothesis. In the course of this, Boyle presented a subtler version of the mechanical philosophy than had his predecessors, not only invoking the size and shape of particles of matter and the motion to which they were subjected, but also stressing the distinctive texture of the bodies that they comprised; his ideas on such subjects had a significant influence on both Isaac Newton and John Locke.

Boyle took it for granted that mechanical explanations were to be preferred wherever possible, but his version of the mechanical philosophy was by no means a simplistic one. He saw no problem in accepting 'subordinate Causes', below 'the most General causes of things', such as the idea that air had a 'spring' in it; he was also receptive to the idea that the corpuscles of matter might have chemical, as against purely mechanical, characteristics; and he even considered the possibility that the Universe might contain 'cosmical qualities' that transcended purely mechanical laws. In addition, he took very seriously the claims of alchemists that they could convert base metals into gold and create powerful medicines by manipulating chemical substances. Though in his famous 1661 book *The Sceptical Chymist*, he was critical of low-level 'chymists', whose view of matter he considered as defective as that of the Aristotelians, he professed great respect for 'the true *Adepti*', and for much of his life he sought knowledge about the secrets of alchemy that he believed might contribute to an understanding of the workings of the world. He even carried out overtly alchemical investigations himself, which in a few cases he published.

Boyle's natural and religious philosophy

In addition to his contribution to an understanding of the nature of matter, Boyle was more widely curious about every aspect of the natural world; he also presented a programmatic view of science that was highly influential. One of his most popular books, entitled *The Usefulness of Experimental Natural Philosophy*, published in two parts in 1663 and 1671, followed the influential example of Francis Bacon earlier in the century in arguing for the potential application of scientific knowledge to every area of human life, from agricultural improvements to industrial and maritime inventions. Science, Boyle thought, would be failing in its potential 'if I thought it could onely teach a Man to discourse of Nature, but not at all to master Her'. The work lay particular stress on the value of scientific findings to medicine, and Boyle was deeply concerned about the amelioration of health, even writing a polemical treatise in which he criticized the standard therapeutical practices of his day, though in the end his wish to avoid antagonizing the medical profession led him to suppress this; instead, his publications on such subjects dealt with topics like the use of specific gravity to detect adulteration in drugs.

Equally important were Boyle's writings on the epistemology of science and on the relations between science and religion. In addition to his writings on experiment, he also reflected with great subtlety on such issues as the appropriate relationship between reason and experience or the relative degrees of certainty that were to be expected in different forms of knowledge. In the latter respect, Boyle was anxious to illustrate the limitations of the human intellect compared with the omniscience of God, and his entire life's work was driven by a deep theism that needs to be taken into account properly to understand him.

Boyle's most popular book was a devotional work rather than a scientific one, and we now know that he experienced a stressful spiritual life that does much to explain his rather driven personality.

He spent hours examining his conscience, and his experimental work could almost be seen as a projection of this soul-searching onto the natural world, as is implied by the quotation at the head of this chapter. Underlying his wish to understand nature was a conviction that thus we come closer to an understanding of God: indeed, it was this sense of the apologetic potential of science that explains Boyle's conversion to experiment in the years around 1650.

Up to a point, Boyle was interested in evidence of a supernatural realm beyond the purely mechanical one that might disarm materialistic views. But no less important was his conviction of the implausibility of any idea that the world had been formed by the random interaction of matter without the supervision of an intelligent designer in the form of God; he was equally hostile to views that saw matter as purposive, which he attacked in his 1686 *Free Enquiry into the Vulgarly Received Notion of Nature*. To understand Boyle we need to place his science in the context of his ideas as a whole.

Antoine-Laurent de Lavoisier
The founder of modern chemistry
(1743–1794)

Nothing is created in the operations of art or of nature,
and it can be taken as an axiom that in every operation
an equal quantity of matter exists both before and
after the operation, that the quality and quantity
of the principles remains the same and that
only changes and modifications occur.

ANTOINE LAVOISIER, *TRAITÉ ÉLÉMENTAIRE DE CHIMIE*
(*ELEMENTARY TREATISE ON CHEMISTRY*), VOL. I, 1789

When Antoine-Laurent de Lavoisier started his studies in chemistry in the 1760s, the discipline was still dominated by Aristotle's theory of the four elements: earth, water, air and fire. By the time of Lavoisier's death three decades later, he had transformed it into a science recognizable in the modern practice of chemistry.

Lavoisier was the son of a wealthy barrister. Born in Paris, he received a formal education at the city's Collège des Quatre-Nations, better known as the Collège Mazarin. After completing the humanities curriculum in 1760, he studied mathematics with the distinguished astronomer Abbé Nicolas Louis de Lacaille. As a result, he said later, he became familiar with the rigour with which mathematicians reason in their treatises. 'They never prove a proposition unless the preceding step has been made clear. Everything is tied together, everything is connected, from the definition of a point to a line and to the most sublime truths of transcendental geometry.'

Leaving the college in 1761, he trained in law to please his father, but also pursued his interest in science. He studied meteorology with Lacaille, continuing even after the latter's death in 1762, and in 1763 published his first scientific report, after observing an *aurora borealis*. He attended courses in chemistry given by the apothecary Charles Louis La Planche at the Society of Pharmacists and three courses taught by the chemist Guillaume-François Rouelle at the King's Garden. He also attended Abbé Jean Nollet's public lectures on electricity and Bernard de Jussieu's lessons on the world of plants. With Jean-Étienne Guettard, a member of the French Academy of Sciences, he investigated mineralogy, geology and chemistry. In 1764, he took his law degree and was accepted as a barrister. But he never practised as a lawyer. He was already preparing himself for entry into the republic of science.

The following year, 1765, he presented his first paper to the Academy of Sciences as a visiting scientist, on 'The analysis of gypsum'. He showed that solid gypsum changed into a powder when heated, and that during this operation a vapour was given off. Having collected the vapour, he found it to be pure water. This was the water of crystallization, and it weighed one-quarter of the weight of the gypsum before the solid was heated. When the water was remixed with the roasted gypsum, crystals formed and the powder became a solid mass again. Thus, the water was the source of the solidity of gypsum. By weighing and measuring ingredients and products during analysis and synthesis, Lavoisier had introduced the 'double demonstration' method that he would use throughout his scientific career. His paper was published in the proceedings of the Academy. A year or so later, he presented a second report on gypsum, showing that it was formed from chalk and sulphuric acid and that its solubility depended on the acid's concentration. Lavoisier insisted that the analysis of mineral substances was important for shedding light on the Earth's past.

Academician of science and tax farmer

In 1766, the Academy announced a competition for the best method of illuminating city streets. Entrants were required to provide 'calculations, physical and chemical experiments, and a theory applied to practice'. Lavoisier carefully studied each kind of lighting, looked at different sizes and types of lamps, and tested oils and candles. In his report, he concluded that olive oil was the best fuel. His essay was awarded a gold medal. Lavoisier eagerly expressed his desire to dedicate his enthusiasm, intelligence and knowledge to serving the state.

In 1767, he made a long field trip to the Vosges Mountains with Guettard to collect data for a great 'Mineralogical Atlas of France'. For more than four months they travelled through the countryside, testing soils, minerals and waters, and sampling agricultural products. The first maps appeared in 1766, eight of which were compiled by Lavoisier. The final edition of the atlas offered twenty-six maps in total, of which sixteen were his work. In 1768, he was elected a member of the Academy.

At the same time, he joined the *Ferme générale* (General Farm), by purchasing a share in this private company entrusted with collecting tax revenue on behalf of the French king. It gathered, for example, taxes on salt, tobacco and alcohol, plus customs duties and the duty on goods entering Paris, by contracting six-year leases with the king's Controller General of Finance and paying a sum in advance to the royal treasury, which was close to 150 million livres in 1770. Whatever amount of money over and above this advance the General Farm collected was its profit. Lavoisier's particular responsibility was for the Tobacco Commission, where his first duty was to fight contraband and fraud among retailers. He would later have much more important responsibilities. His supervisor, Jacques Paulze de Chasteignolles, one of the directors of the French East India Company, was a powerful and rich man. His only daughter, Marie-Anne Pierrette, married Lavoisier in 1771, when he was twenty-eight and she was fourteen years old. Marie brought him a handsome dowry and became her husband's assistant. She learned chemistry, took notes when he did experiments, translated English books of chemistry for him, drew pictures for the books he published, and entertained the many scientists who came to visit him. They had no children.

The four elements

In his first house, at rue des Bons-Enfants in Paris, Lavoisier set up a well-equipped laboratory and began to do research. His work was characterized by its quantitative approach based on constant use of the chemical balance. The law of conservation of mass was already his founding principle. As he would later write in his textbook, 'The whole art of making experiments in chemistry is founded on this principle: we must always suppose an exact equality or equation between the principles of the body examined and those of the products of its analysis.'

He applied the axiom for the first time in print in 1770 in his paper 'On the nature of water'. At this time, many chemists believed that water could be converted into earth because they knew that distilled water, when evaporated in a glass vessel, always left a slight earthy residue. Lavoisier weighed a glass vessel called a pelican and put into it a weighed quantity of water already distilled eight times. The closed vessel was then kept at a temperature of 70° Réaumur (80° Réaumur is the boiling point of water). After 101 days, flakes of powder had appeared in the vessel, but there was no change in the weight of the water. However, the vessel had lost weight, and the loss in weight was about the same as the weight of the flakes. It was clear that the powder had been dissolved from the wall of the glass vessel by the water. Moreover, Lavoisier had established that the water had not been changed to earth.

Of the four Aristotelian elements, Lavoisier now turned from studying earth and water to investigating air and fire. He decided to learn more about what happened to air during burning, or combustion. At the same time he would study a second process using air and fire known as calcination. When metals were heated in air, a powder called a calx formed on the metal's surface; chemists thought that such calcination was a kind of slow combustion. Fire, or rather its main component *phlogiston* – a mysterious substance named by the German chemist Georg Ernst Stahl – was supposedly responsible for both calcination and combustion. Nobody had ever seen phlogiston, but it was thought to be present, in various proportions, in all things that burned. Oil and charcoal were said to be almost pure phlogiston. When substances burned, phlogiston was supposed to be forced out of them, producing fire. During calcination, the same process was thought to produce calx. But Lavoisier was not entirely happy with this explanation, because the weight of the metal plus the calx made from burning the metal in air was more than that of the metal. If phlogiston was ejected from the metal when it was calcined, how could the metal plus calx be heavier? He suspected, on the contrary, that something joined the metal to form the calx and that this something could be air. Thus, the fixing of air would supply the weight gain.

Modern chemistry begins

On 1 November 1772, Lavoisier deposited with the Academy of Sciences a sealed envelope, to be opened in 1773. He had established that sulphur and phosphorus, far from losing weight when burned in air, as commonly supposed, actually gained weight. Furthermore, the calx of lead, known as litharge, when treated with charcoal, lost weight and evolved a large quantity of air. Early in the next year, he wrote out a memorandum of a long series of experiments he intended to make with the purpose of clarifying the role of gases in chemical combination. Among his aims was to discover whether the air that took part in burning and calcining of metals was atmospheric air, or actually a special kind of air – like the so-called 'fixed air' (which we now know as carbon dioxide) discovered by the Scottish chemist Joseph Black in 1754. The answer to this question would bring about a revolution in chemistry.

Joseph Priestley, an English scientist, greatly helped Lavoisier to start this revolution. While in Paris in 1774, he came to dinner at Lavoisier's house and told him he had discovered a new kind of air. Working with calx of mercury, he had found that the calx gave off an air when it turned back into metal, and this type of air had properties entirely different from those of Black's 'fixed air'. Priestley called it 'dephlogisticated air'. But Lavoisier understood the meaning of the phenomenon better than Priestley, without recourse to phlogiston. Since mercury calx, unlike other calxes, did not require charcoal in order to be turned back into metal, the new gas could not have come from the charcoal; it must be from the calx itself. It was a part of atmospheric air.

On 26 April 1775, Lavoisier proudly announced to the Academy of Sciences: 'The principle which combines with metals during their calcination is the purest part of the air. And it follows that fixed air results from the combination of this highly respirable part of the air with charcoal.' Later, he would name the 'purest part' as 'oxygen', from the Greek meaning 'acid producer', since he believed (erroneously) that all acids contain oxygen.

In 1783, the English scientist Henry Cavendish burned hydrogen in closed containers and obtained some water. He correctly concluded that the water resulted from the combustion. But, still convinced that oxygen was lacking phlogiston, he thought that the missing substance had been supplied by the hydrogen. In the presence of several members of the Academy, Lavoisier repeated Cavendish's experiment and demonstrated that hydrogen and oxygen, when burned together, form water. He could therefore assert that water was not an element but a combination of two gases. In 1785, he carried out a great experiment on the analysis and synthesis of water, which both confirmed this discovery and also developed a method for large-scale production of hydrogen. (Soon after, Lavoisier, in arbitrating a competition between the Montgolfier brothers' hot-air balloon and a hydrogen balloon built by another inventor, concluded that hydrogen was a better lifting agent.)

By now, nothing was left of Aristotle's theory of four elements, earth, water, fire and air; and phlogiston did not exist. A new definition of chemical elements was needed and, in 1787, Lavoisier, Claude Berthollet, Antoine François de Fourcroy and Louis Bernard Guyton de Morveau published their new *System of Chemical Nomenclature* that would change our way of conceiving chemistry. Two years later, Lavoisier's textbook *Elementary Treatise on Chemistry* summarized his ideas and offered a list of the known elements.

The chemistry of life

He now studied the physiology of human respiration and came to the conclusion that it was a kind of slow combustion. Thus, oxygen in the air was essential to the chemistry of life. With Pierre Simon Laplace, Lavoisier invented the calorimeter, in order to measure the heat released by an animal, so as to compare it with that released by the combustion of charcoal and thereby determine the animal's expenditure of energy. He then measured animal and human consumption of oxygen, at rest and during work. 'These two memoirs on respiration that I am sending to you are a relatively good beginning to understanding animal physiology', he informed his Italian translator Vincenzo Dandolo. 'But, so far as digestion and the formation of chyle and blood are concerned, everything remains to be done. I have a few ideas and intend to carry out some experiments.'

Lavoisier was among the first to foresee the value of a chemical approach to the physiology of nutrition and to the mechanism of tissue anabolism. In the 1790s, he was ready to start a second scientific revolution, this time in biology. For instance, he had intuited the essential role of the liver in synthesis. He had, indeed, mapped out a conceptual itinerary that would take established science almost a century to open up.

Victim of the terror

But it was not to be. The political revolution in France had started in 1789, and Lavoisier would be drawn into the downfall of the General Farm, which was seen to represent the worst excesses of the Ancien Régime and was hated by revolutionaries. All of his contributions in other domains would be forgotten: the considerable improvement in the production of gunpowder he had obtained as director of the government powder mills; his reform of agriculture, with his demonstration of scientific farming on his model farm at Freschines; his efforts for the adoption of the metric system; his participation in the Advisory Board for Arts and Trades; his reflections on public education; his efforts to save France from bankruptcy, and his memoir on the *Territorial Wealth of the Kingdom of France*, a milestone in the history of economic science, in statistics and in accounting; and finally his work as a commissioner of the national treasury in 1791.

On 28 May 1794, at the height of the Terror, the Jacobins included all of the officials of the General Farm among the alleged opponents of the revolution, arrested twenty-eight of them and organized their cursory trial in front of the Revolutionary Tribunal. All were found guilty and promptly guillotined on the Place de la Revolution. Lavoisier went fourth. 'Only a moment was required to cut off that head,' said the great mathematician Joseph Lagrange, 'and perhaps a century will not be sufficient to produce another like it.'

John Dalton

The development of atomic theory

(1766–1844)

An enquiry into the relative weights of the ultimate particles of bodies is a subject, as far as I know, entirely new; I have lately been prosecuting this enquiry with remarkable success. The principle cannot be entered upon in this paper; but I shall just subjoin the results, as far as they appear to be ascertained by my experiments.

JOHN DALTON, IN A LECTURE FROM 1803

The history of modern atomic theory begins with the unlikely figure of a young Quaker schoolmaster named John Dalton. In a paper read to the Manchester Literary and Philosophical Society in 1803 (from which the above epigraph is taken), Dalton proposed relative atomic weights for several of the most important chemical elements known in his day. Within a decade several leading chemists had adopted Dalton's atomic theory in one version or another, and within a generation the science of chemistry was inseparable from ideas about atoms.

Dalton's father was a poor country weaver who lived near Cockermouth, Cumberland, in the far north-western part of England. His early years were spent working on the family's small farm; but he also pursued an ambitious programme of self-education, aided by mentoring from prominent local Quakers. In fact, the Cumbrian Quakers were notable, even among their co-religionists, for the high value they placed on education and intellectual pursuits. At the age of twelve, Dalton began to teach in the village school, and three years later he joined his older brother in running a boarding school in the nearby town of Kendal. In his spare time, he continued his own education, studying classical and modern languages, mathematics and the natural sciences.

Dalton's favorite pursuit at this time was meteorology, in which he became highly expert, and which he continued to investigate avidly for the rest of his life. In 1793 he published his first book, *Meteorological Observations and Essays*, and in the same year accepted a position as tutor in natural philosophy at the New College in Manchester. When in 1800 that 'dissenters' academy' suffered financial troubles and was unable to pay salaries, Dalton resigned from his post, but he remained in Manchester, supporting himself by giving private lessons on mathematics and chemistry. About the same time he left the New College, he was elected secretary of the Literary and Philosophical Society, in whose lodgings he was given a private study and laboratory.

Dalton lived a quiet life in this bustling and rising English industrial city. He never married, but had a number of close friends, who deeply admired his gentle personality, Quaker 'plainness' and philosophical mind. Although he had not the benefit of higher mathematics, he was exceedingly clever with figures and mathematical concepts, and instinctively applied them to nature. He was also blessed with a vivid scientific imagination. Plain-spoken, totally without pretence or affectation, Dalton quietly pursued his investigations with intellectual courage and understated brilliance.

Weighty matters

From his scientific interest in the atmosphere, Dalton was led to a more general study of the nature of mixed gases, and of gases dissolved in liquids. He became convinced that the only way fully to understand these matters was first to deduce how heavy were the ultimate particles of substances. It was impossible to weigh the atoms of the various elements directly, for they were immeasurably and indeterminably tiny. However, he thought that one might be able to devise a means to infer what were their *relative* weights. For this purpose he arbitrarily assigned the lightest atom, that of hydrogen, the weight of 1, and sought to determine what were the weights of the atoms of each of the other chemical elements in proportion to this weight.

The first step in Dalton's ingenious method was to imagine how a simple compound such as water might be formed, as viewed on the invisible level of ultimate particles. He knew that liquid water was formed from the gaseous elements hydrogen and oxygen, but what would a single molecule of this substance look like? The most likely answer, he thought, was that a single atom of oxygen bonded with a single atom of hydrogen to make water. In today's language, then, Dalton assumed that the formula of water was HO. Dalton's second step was to analyse the compound (or make use of the analyses of other chemists). Contemporary analyses of water indicated that it was about seven-eighths oxygen and one-eighth hydrogen by weight. It was therefore clear that the oxygen atom must also be seven-eighths of the weight of the water molecule. In brief, if the hydrogen atom H is assumed equal to 1 weight unit, and if the water molecule is HO, and if water is seven-eighths oxygen, then $O = 7$. (Today we know that this proportion is actually about eight-ninths.) He applied a similar process to compounds of carbon, nitrogen, sulphur and phosphorus. These were the six atomic weights that appeared in his first paper on the subject, read in October 1803. But as the epigraph suggests, he said nothing to his audience about how he had arrived at these figures. The reason we know this is that we have the evidence of his laboratory notebook. (The actual notebook was destroyed in an air raid in 1944, but crucial pages from it had been published in 1896 in photostat.) The first atomistic calculations in this document appear in an entry dated 6 September 1803, and he continued to pursue these ideas in the months that followed.

There were weaknesses to Dalton's procedure, most obviously the fact that the only way to begin the process was by guessing how many atoms of each element were present in the molecules of these simple substances. This was surely at least one reason why Dalton hesitated at first to reveal the details of his techniques. The first publication of those details, properly credited to Dalton, appeared in 1807 in the chemistry textbook of Thomas Thomson, a friend in whom Dalton had confided. Dalton finally published his own treatment of the theory in the pages of his *New System of Chemical Philosophy*, which was published in 1808–10. In it he set out his thesis that each element is composed of atoms of a single, unique type. All atoms of an element are identical and have the same mass, and atoms of each element are different from one another, with different masses. While they cannot be altered or destroyed, atoms of different elements can be combined in fixed ratios to form compounds of varying complexity. Derived from empirical experiment and analysis, it was the first truly scientific atomic theory.

Some chemists refused to credit Dalton's work, arguing that it was founded on mere hypothesis. What warrant, these critics asked, did Dalton have to assume that the molecule of water really was HO, and not H_2O, HO_2, or any other possibility? Dalton, and others who defended him, conceded that it had been necessary to blindly propose certain molecular formulae, but pointed out that the deduced atomic weights had been cross-checked for accuracy by deriving them from more than one kind of formula. Furthermore, the existence of certain consistent numerical regularities (such as precisely integral multiple proportions of compounds formed from the same two elements) seemed to guarantee that chemical substances really did derive from the combination of atoms forming into molecules in small integral numbers.

The molecular formula of water

In the first few years after Dalton's ideas were published, several other chemists offered differing versions of the atomic theory. Some, such as fellow Englishman Humphry Davy and the Swede Jöns Jacob Berzelius, thought it more likely that the water molecule had two hydrogen atoms rather than just one, for the two gases combine to form water in exactly a 2-to-1 ratio by volume; for them, water was H_2O. But this meant that for these scientists the atomic weight of oxygen had to be 16 times that of hydrogen. There were other disagreements, as well, and the history of atomic theory in the first half of the 19th century is a complex and controverted affair. But despite these complexities, there is no question that Dalton's atomic theory transformed the science. Not only was it possible now to represent elements and compounds in a convenient and beautifully clear shorthand form, but one could actually understand chemical reactions in ways that had never been possible before. Even with its continuing weaknesses, the theory had become a powerful tool for discovery.

Throughout all the controversies, Dalton remained imperturbable, confident in the accuracy of his own atomic weights. He also stubbornly retained his curious ideographic formulae, where atoms were represented by circles of various kinds, even though no one else ever adopted his system of notation. He was sometimes underestimated by his contemporaries, due to his humble personal circumstances. After all, he came from a poor family, he lacked the education and religious affiliation thought to be appropriate to a leading European intellectual, and he had the manners and speech patterns of a provincial northern rustic. To make matters even worse, Dalton failed to keep fully abreast of the rapid advance of the science in the 1820s and 1830s.

But as time passed, Dalton's real merits became ever clearer to the European scientific community. In 1822, on his only trip abroad, he travelled to Paris and was received with the greatest distinction by a galaxy of scientific celebrities, including Laplace, Berthollet, Gay-Lussac, Cuvier and Humboldt. He was made a corresponding member of the French Academy of Sciences, a signal honour. Four years later he received the first Royal Medal, conferred by the Royal Society of London, of which he was a fellow. In 1833 the British government presented him with a life pension of £150 a year; the amount was doubled four years later. And when Dalton died in 1844, his body lay in state in Manchester Town Hall, while 40,000 people filed past to pay their last respects; the funeral cortege the next day was a mile long.

Dmitri Mendeleev

The creator of the Periodic Table

(1834–1907)

The essence, the nature of elements is expressed in their weight, i.e., in the mass of the substance entering into the reaction.... The physical and chemical properties of elements, appearing in the properties of the simple and complex bodies they form, stand in a periodic dependence ... on their atomic weight.

DMITRI MENDELEEV, *PRINCIPLES OF CHEMISTRY*, 1870

Nearly all science students – even beginners – are familiar with the Periodic Table of the chemical elements. If they give the Periodic Table more than cursory thought, they likely consider it to be almost a self-evident system: what other way to order the elements is there than by atomic weight? However, the origin of the Periodic Table was anything but simple; it required synthesizing a huge amount of fragmentary and often erroneous chemical and physical data into a self-consistent system. For this reason, many scholars prefer to use 'Periodic Law' to indicate the deep-seated nature of the web of relationships that encompass the elements when arranged into the Periodic Table. A number of scientists were groping towards a solution to the problem of arranging the elements into some type of table in the 1860s, but most scholars consider the work done by Dmitri Ivanovich Mendeleev, announced in 1869, as the first successful system, even though he would need several additional years to perfect his version.

Mendeleev was born in the small town of Tobolsk in western Siberia, Russia. His father was the director of the local gymnasium, but in 1834, the year of Dmitri's birth, he had to retire on an inadequate pension for health reasons. This forced the family to rely on Mendeleev's mother, who came from an old Siberian merchant family and had inherited a glass factory near Tobolsk, which she now had to operate for the family's sustenance. However, the family's financial situation steadily declined. When Dmitri graduated from the gymnasium in 1849, following the death of his father in 1847 and the destruction by fire of the glass factory in 1848, his mother decided to accompany Dmitri, first to Moscow and then to St Petersburg in an unsuccessful quest to gain him admission to a university. Fortunately for Mendeleev, in 1850 he was finally able to enrol at the St Petersburg Main Pedagogical Institute, the institution his father had graduated from decades previously. His mother died shortly after Mendeleev began his studies there, but he managed to thrive at the Institute and graduated in 1855. After briefly teaching in secondary schools in the south of Russia, he returned to St Petersburg and began studying towards an advanced degree in chemistry.

Mendeleev's early scientific work gained him an extensive knowledge of the chemical properties of the elements and various compounds. His first published work examined the relationship of crystals and their chemical composition, while his master's dissertation considered whether the specific volumes of compounds could be related to their chemical composition or crystallographic form.

In 1859, Mendeleev went abroad on an extended study trip, supported by a government stipend. He spent much of his time in Heidelberg, Germany, conducting original research for his doctoral degree, although he also travelled around Europe. In 1860, he attended the first International Chemical Congress at Karlsruhe, which helped to standardize various chemical concepts such as atomic weights and valence. The conference had a profound impact on Mendeleev's thinking, and it helped to produce or ratify the conditions (in particular, standardized atomic weights) that would prove to be important for the later development of the Periodic Law. It also stimulated other scientists to begin developing schemes for ordering the elements, and several different tables of elements were proposed in 1860s, notably by Lothar Meyer and John Newlands.

On his return to Russia in 1861, Mendeleev taught chemistry at various educational institutions and slowly worked on his doctoral dissertation. In addition, he published on various topics in chemistry. After receiving his doctorate in 1865, Mendeleev became a professor at St Petersburg State University, the country's most prestigious university.

Systematizing the elements

In 1867, dissatisfied with the available introductory chemistry textbooks, Mendeleev decided to write one himself. This turned out to be the decisive step in the discovery of the Periodic Law. The new theory emerged in Mendeleev's process of organizing a large mass of chemical data into a form suitable for convenient and useful pedagogical application. Textbooks at this time often discussed the elements in dictionary style, dividing them into only very general categories, such as metals and non-metals. Mendeleev searched for a more suitable method. He began his *Principles of Chemistry* with extensive discussions of basic definitions, as well as laboratory experiments for students to conduct. He moved on to some of the most prevalent compounds and elements, including salt, oxygen, carbon, nitrogen, and hydrogen. At this point, probably in late 1868 or early 1869, he realized that he would need some other organizational method for the rest of the elements. He now hit upon using atomic weights as the primary quality for each element, which soon led him to the idea of the periodicity of the elements. He quickly wrote up his preliminary conclusions and published them in a Russian journal following a brief presentation at a meeting of the Russian Chemical Society. While it is commonly believed that Mendeleev discovered the Periodic Law in the course of one single day (17 February 1869) that included him receiving inspiration from a dream, it seems far more likely that the Periodic Law was the result of an extended period of thought and contemplation while writing a textbook, rather than of one single spark of inspiration.

Mendeleev had developed the essence of the system, but he still needed to solidify it with detailed chemical and physical data showing the periodic properties of the elements. For nearly two years after his 1869 publication, Mendeleev laboured to buttress his initial insight with a vast array of chemical and physical data derived from his own experimentation and from a thorough search of the scientific literature. He was searching for a 'natural system', in which the properties of each element would be related in a periodic fashion to those surrounding it in the table. By the end of 1871, Mendeleev was confident enough to publish a long paper summarizing his conclusions in a prominent German scientific journal. He left several spaces unfilled in his new Periodic Table and predicted a variety of the chemical and physical properties for the unknown elements that would fill these gaps.

Newly discovered elements

Mendeleev's initial publications on the Periodic Law attracted little attention from other scientists, except from the handful working towards the same goal. This indifference began to change, however, in the latter half of the 1870s and especially in the 1880s. The main reason was the discovery of several new elements, with properties closely matching the unknown elements predicted by Mendeleev. In 1875, the new element gallium was discovered by the French chemist Paul Émile Lecoq de Boisbaudran. Immediately, Mendeleev noted that gallium's properties closely matched those of one of his predicted elements. In 1879, a Swedish chemist, Lars Fredrik Nilson, discovered scandium, and again commented on its close match with Mendeleev's predictions. Numerous scientists also began remarking on how closely his Periodic Table fitted the properties of the existing elements, in addition to the newly discovered ones. In 1886, when germanium was discovered by the German chemist Clemens Alexander Winkler and once again was found to fit Mendeleev's predictions, the Periodic Law was well on its way to becoming a widely accepted principle of science. However, during the years after his proposal of the Periodic Law, Mendeleev fought bitter priority disputes with some of the other scientists, especially Meyer. Mendeleev, who had a fiery and unyielding personality and alienated many of those with whom he came into close contact, likely prevailed as the main discoverer in the eyes of most scientists primarily due to his forceful self-advocacy.

Mendeleev led a distinguished career after the discovery of the Periodic Law. Besides continuing his teaching and scientific research activities, he actively consulted for the Russian government and private businesses on a wide range of economic matters, and ended his professional life as the director of the Central Bureau of Weights and Measures. He became an icon of Russian science and was lionized throughout the Russian Empire as the leading example of Russia's scientific prowess.

August Kekulé

Carbon chains, the benzene ring and chemical structures

(1829–1896)

I turned my chair toward the fireplace and began to doze.
Again the atoms fluttered before my eyes…. But look, what
was that? One of the [figures] had seized its own tail, and
the thing whirled mockingly before my eyes. I awoke as by
a stroke of lightning, and this time, too, I spent the rest of
the night working out the consequences of the hypothesis.

AUGUST KEKULÉ, IN A PUBLIC SPEECH, 1890

Friedrich August Kekulé was the principal creator of the theory of chemical structure, a way of tracing the detailed architecture of complex molecules. One of the most dramatic applications of this theory was to the category known as 'aromatic' substances. This new understanding catalysed the burgeoning chemical industry in the late 19th century.

Kekulé was born and grew up in Darmstadt, the capital of the Grand Duchy of Hesse, a small independent country in central Germany. His father, a member of the Grand Duke's cabinet, wanted his son to be an architect, and August dutifully began to study the subject at the country's small university in Giessen, some miles to the north. But at that time on the university's faculty there was a world-famous celebrity in quite a different field, and after attending Justus von Liebig's course, August became passionately attached to chemistry.

As Kekulé was approaching the end of his studies, Liebig advised him to seek postdoctoral training, partly because there were virtually no available positions for chemists. In the end, Kekulé completed three study trips, the first in Paris, the second in Chur, Switzerland, and the last in London. He then qualified as lecturer at the University of Heidelberg. After two and a half years in that position, in the fall of 1858 he was finally hired as professor of chemistry at the University of Ghent, in francophone Belgium. Nine years later, by then one of the most famous chemists in Europe, he was called to the University of Bonn, and he spent the rest of his career in that Rhineland city. Delighted as he was to return to his homeland, Kekulé's private life was not happy, for his beautiful young wife had died delivering their first child, and his second marriage was not fortunate.

A visionary theorist

In the 1840s chemists were beginning to develop possible ways to determine the arrangements of atoms within molecules, but there was also a great deal of controversy and confusion on this subject. As a young theoretical chemist, Kekulé found himself in the forefront of these debates, and he had the advantage of intensive profes-sional immersion not just in Germany, but also in France, England, Switzerland and Belgium. Several European scientists were just then beginning to develop ideas concerning the valences of atoms – for instance, that a hydrogen atom can form a bond to just one other atom, an oxygen atom to two, nitrogen to three, and carbon, perhaps, to four other atoms.

According to a story Kekulé told as an old man, it was during his third postdoctoral stint, while daydreaming on the upper deck of a horse-drawn London omnibus on a summer evening in 1855, that a molecular vision appeared to him. Having arrived home, he wrote down details of how one might schematically dissect many molecules right down to their individual atoms – the theory of chemical struc-ture. The key was not just that each carbon atom can bond to four other atoms, but also that they can bond to each other, forming linear chains of carbon atoms. He published this theory three years later; it soon became the central doctrine of chemical theory, a priceless guide for chemical analysis and synthesis.

Kekulé told a second story on the same occasion, about an evening in 1862 when, while half-dozing before the fireplace in his apartment in Ghent, he saw a vision of a snake grabbing its own tail. This time the molecular dance in his mind's eye gave him the idea that benzene – the basis of all so-called 'aromatic' substances – might have a molecule shaped like a ring rather than a straight line. This notion became the heart of his benzene theory, first published in 1865. Fortuitously, the synthetic dye industry was just then becoming big business, especially in Germany, and virtually all of the new dyes were derivatives of benzene. In fact, it was not just dyes, but also pharmaceuticals, food chemicals, munitions, plastics and synthetic substances of all kinds that were built on an 'aromatic' foundation. Kekulé's new and better scientific understanding of these substances worked like a master key to productive research, and played a major role in the explosive growth of several new chemical industries.

Kekulé was one of the most creative scientists of the 19th century. His extraordinary energy and good humour, sparkling personality and scientific charisma attracted an international crowd of students, friends and admirers. But it was his countrymen who were the most indebted to him, for organic chemistry in Germany, assisted largely by his ideas, rose to a dominant international position towards the end of his life. Those ideas still form the foundation of organic chemistry today.

Dorothy Crowfoot Hodgkin

The structure of complex biological molecules

(1910–1994)

*I should not like to leave an impression that all structural
problems can be settled by X-ray analysis or that all
crystal structures are easy to solve. I seem to have
spent much more of my life not solving structures
than solving them.*

DOROTHY CROWFOOT HODGKIN, NOBEL LECTURE, 1964

Dorothy Crowfoot Hodgkin dedicated her working life to finding the molecular structures of medically important natural chemicals, such as antibiotics, vitamins and proteins. The only British woman ever to have won a Nobel Prize in science, she attracted widespread admiration for her devotion to the cause of world peace, and for her efforts to promote science and education in the developing world. Long before it was commonplace for women to continue to work after marriage, she supported her husband in his own demanding career and brought up three children, while pursuing her pioneering scientific research.

Born Dorothy Crowfoot, the eldest of four daughters of a British colonial administrator and archaeologist, Hodgkin first grew crystals in a small chemistry class when she was ten years old. She was 'captured for life', and immediately started to do experiments at home. In 1928 she was accepted to read chemistry at Somerville College, one of the colleges for women at Oxford University. She obtained a first-class degree, and in 1932 went to Cambridge to study for a PhD with John Desmond Bernal. A brilliant crystallographer and campaigning left-wing thinker, Bernal had begun to work on biological molecules. Hodgkin became his closest assistant as well as sharing his passionate socialist principles.

The activity of the human body's natural chemicals depends on the way the tens, hundreds or even thousands of atoms in each molecule are connected together in a precise three-dimensional arrangement. By firing a beam of X-rays through a pure crystal of a substance and measuring the positions and intensities of the scattered rays, it is possible to reconstruct the positions of the atoms relative to one another. This technique, X-ray crystallography, was first demonstrated by William and Lawrence Bragg in 1912. Bernal and Hodgkin were the first to apply it to complex biological molecules such as the digestive enzyme pepsin.

In 1934, Hodgkin returned to Oxford University. Somerville College had given her a research fellowship, and the professor of organic chemistry there, Robert Robinson, obtained a grant for her to set up her own X-ray laboratory within the university museum. Almost immediately, she began work on the protein hormone insulin, but the molecule was too large, and the apparatus too primitive, for an immediate solution. It would take her over three decades to discover its complex structure.

Soon after her return to Oxford she met Thomas Hodgkin, and they married in December 1937. Their children were born between 1938 and 1946, during which period she continued with her research. Penicillin had been isolated by researchers working in Oxford's Dunn School of Pathology, and was successfully tested on human patients for the first time in 1941. In a time of war, it was a matter of priority to analyse the arrangement of its two dozen or so atoms to help with mass production of the drug. By VE Day in May 1945 Hodgkin had succeeded, resolving a dispute between the chemists and demonstrating that X-ray crystallography could reveal structures even when the chemical formula was uncertain.

As her fame grew, she attracted students and more senior colleagues from all over the world. Her next major success was vitamin B12, used to treat pernicious anaemia, which she solved in 1955. After receiving many other honours she was awarded the Nobel Prize in Chemistry in 1964. Her crowning achievement was the discovery of the structure of insulin, a molecule with thousands of atoms, which she finally identified in 1969.

A passionate campaigner for peace

As a Nobel laureate, she realized that her support could be valuable to other causes. In 1975 she became president of the Pugwash Conferences on Science and World Affairs, which brought together scientists from East and West to campaign against nuclear weapons; she supported organizations fighting for peace in Vietnam; as chancellor of the University of Bristol from 1971, she campaigned against cuts in university budgets; she made many visits to China, India and other developing countries, encouraging the exchange of students and scientists with the better-resourced institutions of the developed world; and she urged Prime Minister Margaret Thatcher, who had been her student at Somerville, to open a dialogue with the Soviet Union.

Despite her great eminence, Dorothy Hodgkin was gentle, modest and quietly spoken. She encouraged many women to continue with careers in crystallography, partly by her example and partly through the direct help and support she gave them. She showed tremendous courage throughout her life, not only in forging a successful career in a new field of research, but in coping with the increasing pain of arthritis, which afflicted her from the age of twenty-eight. In the summer of 1993, although confined to a wheelchair, she made a final visit to Beijing for the International Congress of Crystallography. Her friends and colleagues from all over the world were thrilled and moved to see her there, dedicated to the last to sharing in a great scientific adventure.

Chandrasekhar Venkata Raman
Molecular physicist and theorist of light
(1888–1970)

The real inspiration of science, at least to me,
has been essentially the love of nature.

CHANDRASEKHAR VENKATA RAMAN, *WHY THE SKY IS BLUE*, 1968

The Indian physicist Chandrasekhar Venkata Raman discovered that the scattering of light by the molecules in a gas, liquid or solid causes a change in its wavelength, an effect known as Raman scattering. The spectra arising from this scattering, called Raman spectra, are used to identify and analyse the molecular structure of materials. For this discovery, Raman was knighted and awarded the Nobel Prize in Physics in 1930, becoming the first Asian to win a science Nobel Prize.

Raman was born in Thiruvanaikaval, a village near Tiruchirappalli in the south Indian state of Tamil Nadu. His father, Chandrasekhar Iyer, was a professor of physics and mathematics. Raman went to Presidency College in Madras for his graduate studies. There he wrote his first published research paper, 'Unsymmetrical diffraction', on optics, which appeared in the *Philosophical Magazine*. During those colonial days it was not possible to pursue a scientific research career in India without a degree from a British university. So Raman took the competitive examination for the Audit and Account service of the Government of India, and was ranked first. He was appointed assistant accountant general in 1907, and remained an accountant in Calcutta for ten years. Soon after his arrival, Raman came across the Indian Association for the Cultivation of Science, which had been founded by Mahendralal Sircar in 1876 to provide a forum, managed by Indians, for lecture demonstrations. Raman started working there in his spare time, even though the facilities were rather meagre. His earliest significant work was on the physics of bowed strings covering instruments of the violin family, which extended the description of the fundamental mode of vibration given earlier by Hermann von Helmholtz to more complex modes.

Raman's work at the association brought him to the attention of the builder of Calcutta University, Sir Asutosh Mukherji, who offered him the newly endowed Palit Professorship in Physics. Raman accepted in 1917, even though it meant the sacrifice of a lucrative civil-service career and a drastic cut in salary.

Raman made his first trip abroad, to England, in 1921, for a science conference of representatives from universities across the British empire. Returning to India by sea, he was struck by the intense blue colour of the Mediterranean. The physicist Lord Rayleigh had attributed the blue to the sea's reflection of the blue colour of the sky, which is caused by the elastic scattering of sunlight in the atmosphere, known as Rayleigh scattering. But an observation of the sea surface through a Nicol prism, a device that polarizes light, held at the 53° angle needed to eliminate reflected sunlight (the 'Brewster angle'), ruled this out. After making experiments in Calcutta, Raman concluded instead that the blue colour is due to light scattering by water molecules, just as the colour of the sky is the result of scattering by air molecules. This conclusion resulted in Raman's small volume *Molecular Diffraction of Light*, published in 1922, and ultimately led to an intense period of experiments and his discovery of an effect that would bear his name.

Shedding light on scattered light

Raman scattering was first noticed in Raman's laboratory around 1923, and published in 1928 in the *Indian Journal of Physics*, which he had established. Light incident on a transparent material was found, after it had been scattered, to have a weaker secondary component with a modified frequency (that is, energy level), in addition to its primary Rayleigh scattering component, which has an unchanged frequency from the incident light particles. Initially, Raman scattering was thought to be due to fluorescence, but Raman ruled out this explanation in experiments with K. S. Krishnan, by showing that the scattered light was strongly polarized. Early in 1928, Raman realized that his observed secondary radiation was the optical analogue of the X-ray scattering that had been discovered by Arthur Compton in 1923, in which X-rays passing through matter were scattered and emerged with a longer wavelength.

In the Compton effect, X-ray radiation behaves as quantized particles (photons) undergoing elastic collision with electrons in matter. The effect was decisive evidence for the existence of such quanta, with energy and momentum proportional to their frequency. In the Raman effect, visible light behaves as quantized particles that undergo inelastic collision with molecules. Raman scattering has either a lower, or a higher, frequency than the incident radiation, depending on whether the light quanta impart energy to, or absorb energy from, the molecules. The theory of the effect was envisaged by Werner Heisenberg and Hendrik Kramers in 1925 in their work on the quantum theory of dispersion. The Raman effect was thus seen to provide strong evidence for the quantized nature of light.

The main importance of the Raman effect was that it gave rise to a powerful technique for studying molecular structure and energy levels. The shift in frequency in Raman spectra between the incident and the secondary radiation relates directly to the difference between the initial and the final molecular energy levels, and therefore the Raman effect can be used to identify specific molecules and the chemical bonds. To begin with, the available information involved mostly the rotational and vibrational levels of molecules. These were previously available only from infrared spectra, which were difficult to obtain;

Raman optical spectra made such information more convenient and accessible. With the invention of the laser in the 1960s, Raman spectroscopy became even more advanced and accurate, enabling the technique to be used in the microscopic examination and measurement of materials and their properties. Today it serves many diverse functions in a wide range of fields, from its use in medicine for the real-time monitoring of anaesthetic gases during surgery, and its employment in the conservation of historical artefacts, to its application by law-enforcement and security services to identify drugs, detect explosives and trace forensic evidence.

Raman left Calcutta in 1933 to join the Indian Institute of Science at Bangalore as its first Indian director. Both in Calcutta and in Bangalore he trained a large number of students who went on to occupy important positions. Although he resigned the directorship of the institute after four years, he continued as professor of physics until his retirement in 1948, when he founded the Raman Research Institute, where he worked on the optics of minerals and the physiology of vision. His most notable contribution from this period was the Raman-Nath theory of diffraction of light by ultrasonic waves. During the 1940s, Raman clashed with Max Born over the Born-Van Karman theory of lattice vibrations. This theory predicted a quasi-continuum for Raman spectra, whereas Raman found prominent discrete features for the spectra of diamond. The resolution of this controversy was provided by others in 1953: the observed discrete features were attributed to the singularities present for some of the normal modes in the quasi-continuum.

Raman was a naturalist at heart, fascinated by the beauty of nature – whether in the colours of the sea or of minerals. He took delight in sounds, which led to his work on musical instruments and in whispering galleries. Throughout his science, he celebrated natural beauty by investigating its physics.

INSIDE THE ATOM

Towards the end of the 19th century, when Einstein was still a school-boy, many physicists suspected that their subject had reached a kind of terminus, with no more major discoveries in prospect. In particular, the atom – assuming that it existed – was regarded as indivisible, containing no major secrets. Then, over a mere two decades starting in 1895, X-rays were observed by Wilhelm Röntgen; radioactivity in uranium, radium and some other elements was discovered by Henri Becquerel and the Curies; the negatively charged electron was detected by J. J. Thomson; and the radioactive transmutation of elements, isotopes, alpha and beta particles, the dense atomic nucleus and the positively charged nuclear particle later known as the proton were discovered by Ernest Rutherford, Frederick Soddy and collaborators. In the same period, Max Planck developed the quantum theory; Einstein created his theories of special and general relativity; and Rutherford's student Niels Bohr visualized the solar-system model of the atom, in which electrons were imagined to revolve around the nucleus like planets around the Sun – their orbits fixed according to discrete electronic energy levels dictated by the quantum theory.

During the 1920s and 1930s, the structure of the atom became more and more complicated. As the quantum revolution devel-oped in the hands of Bohr, Max Born, Louis de Broglie, Paul Dirac, Werner Heisenberg, Wolfgang Pauli, Erwin Schrödinger and Richard Feynman (to mention only certain key physicists), it became clear that subatomic particles, such as the electron, should be regarded not only as particles fixed in their separate orbits but also as waves. In the quantum-wave mechanical model of the atom, Bohr's neat 'solar system' was replaced by a less easily visualized wave function for the orbiting electrons, which defined their position in terms of probability rather than certainty. During the 1930s, the chemist Linus Pauling was able to use this newly discovered quantum physics for his work on the nature of chemical bonding, which involves the sharing of electrons by atoms through ionization or covalency, leading to a picture of crystals and molecules that revolutionized chemistry too.

Over the coming decades, new subatomic particles, some nuclear, were discovered or predicted to exist. In 1932, James Chadwick, working under Rutherford, discovered a second nuclear particle: the neutron, with a mass similar to the proton but no electric charge. But it was not obvious what kind of force was holding the nucleus together, given that two or more positively charged protons must repel each other electromagnetically at close range. Hideki Yukawa therefore postulated nuclear particles named mesons, intermediate in mass between the light electron and the heavy proton, capable of transmitting the strong interaction that binds the nucleus together; the first meson, named a pion, was observed in 1947.

Outside the nucleus, Dirac, employing quantum mechanics, special relativity and the new concept of electron spin, predicted the existence of an 'anti-electron' with the same mass as an electron but carrying an equal but positive charge; this was observed in 1932 and named the positron. At the same time, Pauli predicted the existence of an electrically neutral, non-nuclear, particle with a mass close to zero and half-integral spin; dubbed a neutrino by Enrico Fermi, this was finally observed in 1956. With the development of particle accelerators and detectors from the 1950s, many more subatomic particles were discovered, leading to the creation of what has become known as the Standard Model of particle physics since the 1970s. However, subatomic particles have continued to be discovered, most recently at the Large Hadron Collider at the CERN laboratory in Geneva, and the current understanding of the atom remains in a state of flux.

Einstein – a severe critic of quantum theory from 1925 and a sceptic about elementary particle physics in his later years – played little part in the above developments. But as usual his words are worth heeding. Writing in 1938 in *The Evolution of Physics: The Growth of Ideas from the Early Concepts to Relativity and Quanta*, he commented: 'Science is not and never will be a closed book. Every important advance brings new questions. Every development reveals, in the long run, new and deeper difficulties.' How prescient his thoughts have proven to be. In 2011, particle physicists working near Rome announced they had detected neutrinos that had apparently travelled from CERN faster than the speed of light – thereby overturning one of the cornerstones of modern physics, first laid down by Einstein. But soon these physicists located flaws in their equipment setup, and the current model was preserved for now. In 2012, other physicists measured the neutrino travelling at the speed of light, in line with Einstein's theory of relativity. By coincidence, this corrected result was announced at the same time as a striking CERN discovery concerning a long-sought elementary particle. First predicted by the theoretical physicist Peter Higgs back in 1964, and then controversially dubbed the 'God particle' by the leading experimental physicist Leon Lederman, it was now termed the Higgs boson.

Marie Curie and Pierre Curie

Pioneers of radioactivity

(1867–1934 and 1859–1906)

*A great discovery does not leap completely achieved from
the brain of the scientist, as Minerva sprang, all panoplied,
from the head of Jupiter; it is the fruit of accumulated
preliminary work. Between the days of fecund productivity
are inserted days of uncertainty when nothing seems to
succeed, and when even matter itself seems hostile; and
it is then that one must hold out against discouragement.
Thus without ever forsaking his inexhaustible patience,
Pierre Curie used sometimes to say to me: 'It is
nevertheless hard, this life that we have chosen.'*

MARIE CURIE, *PIERRE CURIE*, 1923

The discovery and isolation of radium by Marie Curie, collaborating with her husband Pierre Curie, between 1898 and 1902, is often regarded as a story of elemental simplicity. The heroic legend that has grown up around her name – epitomized by Albert Einstein's comment that 'Marie Curie is, of all celebrated beings, the only one whom fame has not corrupted', and symbolized by today's charity Marie Curie Cancer Care – has emphasized this view. But in fact, it was a complex interaction between physics and chemistry, involving exact observation, subtle thinking, cutting-edge technology, brute-force methods, extreme dedication and good luck, that brought Marie the extraordinary accolade of two Nobel Prizes: the first in physics in 1903, the second in chemistry in 1911.

Driven to succeed

Her upbringing in Poland was crucial to her success. Warsaw, where Manya Skłodowska (Marie Curie's unmarried Polish name) was born, was under harsh Russian rule throughout her youth. Several of the Skłodowski family took up arms against the regime. Manya's parents were leading members of the unarmed, intellectual resistance, as distinguished school teachers. In general, the Russian educators treated their Polish pupils as 'enemies', Curie recalled in her autobiography. The result was a Polish patriotic fervour that inculcated in Manya the austere combination of determination to succeed, passion for knowledge and moral conviction that would define her career. Her mother died of tuberculosis when she was only ten, but 'thanks to her father she lived in an intellectual atmosphere of rare quality known to few girls of her age', wrote Eve Curie, Marie's second daughter.

Having finished school with a gold medal, Manya faced the lack of higher educational options for women, and the need to make a living. Eventually she had to sign on as a governess for three-and-a-half years. The money she earned went to help her elder sister train as a doctor in Paris, on the understanding that once her sister was established, Manya would follow her. In 1891, she enrolled as one of twenty-three women students joining the sciences faculty of the Sorbonne. But at no point in her own accounts of her student life did she mention that hers was a male-dominated academic world. Indeed, Marie Curie never encouraged feminist attempts to cite her as an example to other women.

After less than three years of gruelling study, she ranked first in the examination for the *licence ès sciences*, and second in the exam for the *licence ès mathématiques*. One of her professors introduced her to Pierre Curie, a somewhat older physicist already much respected for his work on piezoelectricity – the electric charge that accumulates in certain solid materials such as crystal, ceramic and bone – and on the effects of temperature on magnetism, in the hope that he would have facilities in his laboratory suitable for her research. The similarities in their background, in Poland and France, were striking. They married in 1895, and began to collaborate scientifically after the birth of their first child Irène (later herself a Nobel laureate) in 1897. Their handwriting alternates throughout their laboratory notebooks over the next few years. Not only was there a constant exchange of ideas between the couple, there was also 'an exchange of energy', said Henri Poincaré, 'a sure remedy for the temporary discouragements faced by every researcher'.

A puzzling phenomenon

Uranium's radiation had been discovered in Paris in 1896 by Henri Becquerel, while investigating Wilhelm Röntgen's recent discovery of X-rays and their luminescent effects on certain minerals. Becquerel tested a number of luminescent minerals by placing them in bright sunshine for several hours, in the form of a thin crystalline layer on top of two sheets of very heavy black paper wrapped around an unexposed photographic plate. His idea was that neither sunlight nor fluorescence would be able to fog the plates because of the paper, but that any 'invisible fluorescence' – in other words, the emission of rays from the mineral similar to X-rays – should be detectable as dark patches on the plate. With uranium salts, he discovered that the paper was not opaque to radiation from the mineral; part of the photographic plate was fogged by a dark silhouette of the mineral layer. He therefore assumed that sunlight had stimulated the emission of invisible rays from the uranium. But then a period of several cloudy days intervened. A disappointed Becquerel held back some of the prepared plates, with their uranium coatings, and stored them in a closed laboratory drawer. When he developed them anyway, he famously got a shock. Instead of finding very weak shadows from the uranium layers as he anticipated, 'the silhouettes appeared with great intensity. I thought at once that the action must have been going on in darkness', Becquerel reported. He had discovered radio-activity (for which he would share the 1903 Nobel Prize in Physics with the Curies); but he did not name the phenomenon, and had no explanation for it.

The Curies decided to investigate the new radiation by testing a range of minerals as precisely as possible. Becquerel had shown that radioactivity discharged electrified bodies, as well as affecting photographic plates. To detect such ionizing radiation, Pierre therefore designed an extremely sensitive current-measuring instrument: an electrometer combined with a piezoelectric quartz balance. In essentials, the apparatus consisted of a condenser (ionization chamber); an electrometer to measure differences in electric potential; and a piezoelectric quartz crystal. Piezoelectric crystals have the property of producing a minute electrical polarization across their crystal faces when mechanically stressed. In this case, small weights suspended from the bottom of the crystal produced the polarization. The substance to be tested was spread as a fine powder on the bottom plate of the condenser, which was connected to one pole of a 100-volt battery. The top plate was connected to one terminal of the electrometer, and the other terminal to the top of the quartz crystal. (The bottom of the crystal was earthed, as was the other pole of the battery, thus making a complete electric circuit.)

During operation by Marie, the slow increase in electric charge on both plates caused by the ionization of the air in the condenser by the substance's radiation was counterbalanced by the increase in charge of the quartz crystal generated by gradually adding weights to it. The balance point was detected by the electrometer. This was made of a rotating blade of aluminium suspended from a conducting platinum wire with a small mirror below it; a beam of light falling on the rotating mirror produced a spot of light on a graduated glass scale. When the spot fell on the centre of the scale (regarded as zero), the charges on the top plate of the condenser and on the piezoelectric quartz crystal were exactly equal. The trick was to keep the spot of light in the centre as the experiment proceeded. Remaining as still as possible, with one hand Marie had to add weight after weight to the crystal and with the other hand, start and stop a chronometer, while continually monitoring the movement of the spot of light with her eyes. After time T from the beginning of the experiment, the amount of charge Q on the plate of the condenser was equal to the charge on the crystal. The current flow caused by the radiation was then given by the flow of charge per second – that is, Q divided by T.

In April 1898, Marie, working alone, reported that: 'Two minerals of uranium, pitchblende (a uranium oxide) and chalcolite (uranyl copper phosphate) are much more active than uranium itself. This fact is most remarkable, and suggests that these minerals may contain an element much more active than uranium.' Naturally occurring chalcolite registered a current of 52 millionths of a millionth of an ampere, as opposed to only nine millionths of a millionth of an ampere for artificially prepared chalcolite.

Isolating radium

The obvious next step was to try to isolate the unknown chemical element. Pierre now joined Marie full time, despite his being more a physicist than a chemist. They thought that isolation would require perhaps a few weeks. In the event it took them, and Marie in particular, a few years; and also determined the course of her entire life. With the help of a chemist colleague, the Curies developed a purification method, and produced a substance about 400 times more active than uranium. From first chemical, and later spectroscopic, analysis, it was clear that there were at least two new elements in pitchblende. In July 1898, they named the first element polonium, and in December, the second element radium. The title of their joint paper, 'On a new radioactive substance contained in pitchblende', marked the earliest use of 'radioactive' as a scientific term.

After highly laborious purification from tons of pitchblende, the end product in 1902 was a tenth of a gram of pure radium chloride – a minute one-fiftieth of a teaspoonful. But it was enough for Marie to determine the atomic weight of radium as 225 (very close to today's value 226) and to place radium below barium in Mendeleev's Periodic Table of the elements, in the column of alkaline earth metals. By 1910, working with another chemist, André Debierne (who in 1899 had discovered the element actinium in pitchblende), she had prepared radium as a pure metal. Pure radium became a standard of comparison for other radioactive substances, not least those used in radiotherapy. 'It is not an exaggeration to say today that [the isolation of radium] is the cornerstone on which the entire edifice of radioactivity rests', wrote the physical chemist and Nobel laureate Jean Perrin in 1924 – even though the correct theoretical explanation of radioactivity in terms of the structure of the atomic nucleus would be left to contemporaries of the Curies, notably Ernest Rutherford and Frederick Soddy.

The death of Pierre Curie in a Paris street accident in 1906 was a bitter blow for Marie, which cast a shadow over the rest of her life. But it did nothing to reduce her commitment to science. She was immediately appointed to the professorship Pierre had held, and she became the first woman to teach at the Sorbonne, her old university. In 1914, under her direction, the laboratories of the Radium Institute at the University of Paris were completed, which in due course became a universal centre for nuclear physics and chemistry, where Irène and Frédéric Joliot-Curie would make their discovery of artificial radioactivity in 1934. The medical applications of radioactivity increasingly preoccupied Marie; during World War I, she became head of the French Radiological Service, and herself drove ambulances equipped for X-radiography to the front lines. She died in a sanatorium at Haute-Savoie at the relatively early age of sixty-seven, from leukaemia, undoubtedly caused by her years of prolonged exposure to highly radioactive materials.

Ernest Rutherford

Penetrating the secrets of the atomic nucleus

(1871–1937)

*It is given to few men to achieve immortality, still less
to achieve Olympian rank during their own lifetime.
Lord Rutherford achieved both.*

NEW YORK TIMES OBITUARY OF ERNEST RUTHERFORD, 1937

Ernest Rutherford was the father of nuclear physics. Although best
known for his discovery of the atomic nucleus, which is the kernel
of much of modern physics, he won his Nobel Prize in chemistry
– for discovering the transmutation of the elements. High-energy
particle physics, which studies the nature of matter and its origins in
the Big Bang, and modern nuclear technology – from nuclear power
and weapons to nuclear medicine – are but a few examples of his
legacy. Without argument, it can be said that the depth and breadth
of Rutherford's discoveries in experimental physics were the equal
of Einstein's in theoretical.

Born in New Zealand, he came to England in 1895 almost by
accident. New Zealand's winner of the international scholarship that
year had declined it in order to get married; Rutherford was second
in line and, as they say, the rest is history. Yet even upon arrival in
Cambridge, his future could have turned out very different. In New
Zealand he had been researching in electronic technology and briefly
held the world record for the distance over which electromagnetic
waves could be detected. Ahead of Guglielmo Marconi at that time,
he planned to continue in this field, but that same year, 1895, Wilhelm
Röntgen had discovered X-rays, soon followed by Henri Becquerel's

discovery of radioactivity, and the focus of work in Cambridge, under J. J. Thomson – himself about to discover the electron – had turned to these mysterious novel radiations. According to legend, Thomson sought advice from Lord Kelvin, who opined that there would be no future in radio; with this view in mind, Rutherford was pointed towards radioactivity.

Radioactivity had been discovered by Becquerel in 1896, and named by the Curies in 1898 in the process of discovering radium; however, it was Rutherford who used these breakthroughs as a scientific tool. He used the radiations as a means of bombarding atoms and probing their inner structure. It was with these investigations – first as a research student at the Cavendish Laboratory in Cambridge in the 1890s, then as professor at McGill University in Canada, and later in Manchester – that he cracked the secrets of atomic structure. In 1919, he returned to Cambridge, succeeding his mentor Thomson, as the greatest experimental physicist of the day.

Radioactivity as alchemy

Initially he had been working with Thomson on the ionizing effects of X-rays in gases, and then moved to his first great contribution: measuring the intensity of radiation from uranium. In the course of this, he covered a sample of uranium with sheets of aluminium foil. These absorbed the radiation. As he increased the thickness he found that the intensity of radiation dropped, which is as expected: the radiation is progressively absorbed. However, upon increasing the thickness further, he found that the intensity seemed to stay about the same. Only after adding several more layers of aluminium was he able to tell that the intensity was falling away after all, but much more slowly than before. From this, he deduced that there must be two forms of radiation. One, which was rapidly absorbed, he called alpha; the other, which he called beta, penetrated the foil. Later he found a third, more penetrating form, which he called gamma.

In 1898 he moved to McGill University in Montreal, Canada. The first thing that he did there was to measure the amount of energy being emitted by uranium. This was his great epiphany: the amounts were up to 100 times greater than are produced in any known chemical reaction. With this measurement he had the first glimpse of the latent power deep within the atom.

Thomson had shown in 1897 that atoms contained small electrical particles – electrons. To balance their negative electric charge, there had also to be positive charge within the atom, which implied that atoms had to be complex structures. Rutherford proposed in 1900 that the source of uranium's power was the rearranging of the atomic constituents. As the structure of atoms remained to be explained, this was a remarkable insight for the time, not least because others, notably the Curies, erroneously believed that the energy of radioactivity was external to the atom.

It was around this time that Rutherford became intrigued by the unusual behaviour of radioactivity from the element thorium. The amounts seemed to vary, and were sensitive to breezes in the air. After a series of experiments he decided that thorium emitted a radioactive gas, which was disturbed by air currents. To determine what this gas consisted of required the help of a chemist: Frederick Soddy. They found conclusive proof that it was a new element: radon. With this discovery they had found the first proof that one element – thorium – can turn into another – radon. When Soddy exclaimed 'Rutherford, this is transmutation!', Rutherford replied: 'Don't call it transmutation; they'll have our heads off as alchemists.' This however was what they had achieved; it was indeed alchemy, but it was occurring naturally. Eventually, they showed that thorium first transmutes into radium, which in turn converts into radon, at each stage emitting radiation. In addition to demonstrating the remarkable cascading through the Periodic Table from one element to another, they had shown that radiation is the direct result of transmutation.

The discovery of the atomic nucleus

In 1907 Rutherford moved from McGill to Manchester University. He gathered around him a team, including the young German Hans Geiger, today famous for the Geiger counter. Rutherford and Geiger used a prototype of the eponymous counter to study alpha rays, showing that these consist of particles that are positively charged and nearly 10,000 times more massive than an electron. They are in fact the doubly charged nuclei of helium atoms. In 1908 Rutherford had managed to prove this by collecting lots of alpha particles, neutralizing them with electrons and then exciting a spectrum of the resulting gas. It proved to be identical to that of helium. He announced this in his speech when he was awarded the Nobel Prize that year – not for physics, but for chemistry – for his work with Soddy on transmutations. Rutherford had always rated physics far ahead of chemistry, even alluding to the latter as mere stamp collecting. Nonetheless, he led the good humour about his 'instant transmutation from physicist to chemist'. That the award was for chemistry was correct: Rutherford and Soddy had used physical methods, but it was the field of chemistry that had been revolutionized. Lady Rutherford was assured during the Nobel ceremony that her husband 'would win the prize for physics one day'. However, he never did, which is surprising since his lifelong train of discoveries had hardly begun.

The identification of alpha particles gave conclusive proof that heavy atoms can decay into lighter ones by ejecting tiny atomic fragments. This left an unanswered question of how these pieces – the lightweight negatively charged electrons and positively charged counterparts – were arranged within atoms. Solving this fundamental problem would be his next dramatic contribution.

Rutherford and Geiger built a screen coated with zinc sulphide, which would emit faint flashes when struck by electrically charged particles, such as an alpha particle. During his time at McGill he had noticed that alpha particles seemed to be scattered from their line of flight as they passed through thin sheets of mica. This was a surprise as the alpha particles were moving at 15,000 kilometres per second, about one-twentieth of the speed of light. For them to be deflected at all implied the presence of electric and magnetic forces far greater than anything then known. This gave him the idea that these powerful forces are present within atoms.

Geiger had a young student named Ernest Marsden. In 1909, Rutherford suggested that Marsden look to see if any alpha particles were deflected through very large angles. Marsden used gold leaf rather than mica, and a scintillating screen to detect the scattered alpha particles. To everyone's amazement, Marsden reported that about one in 20,000 alphas bounced right back from whence they had come. This was after hitting a thin gold sheet that was only a few hundred atoms thick. Rutherford famously exclaimed: 'It was as though you had fired a 15 inch shell at a piece of tissue paper and it had bounced back and hit you.'

He spent a year puzzling over this phenomenon, until he realized that it implied that the positive charge in an atom is concentrated in a massive and exceedingly compact central 'nucleus'. It was the repulsion of like charges that was turning the relatively lightweight alpha (the nucleus of a gold atom being some fifty times more massive than an alpha particle) back in its tracks. The size of the nucleus relative to an atom was famously compared to being like a 'fly in a cathedral'.

With his discovery of the atomic nucleus, he had sown the seeds for Niels Bohr, Rutherford's assistant in Manchester, to invent his atomic model in 1913. This led to the popular image of an atom as a miniature 'solar system', with lightweight 'planetary' electrons encircling a nuclear 'sun', yet with the ideas of the infant quantum theory brought to bear. Although the details have subsequently developed, incorporating mature ideas of quantum mechanics and relativity, the essence of this simple picture has survived for a century.

Splitting the atom and the birth of nuclear physics

These experiments had revealed the existence of the atomic nucleus, but had not probed the structure of the nucleus itself. In an electrically neutral atom, the positive electric charge of the nucleus counterbalances the negative charge of the surrounding electrons. Rutherford realized that as the atoms of light elements have fewer electrons than those of heavy atoms, the charge on their nuclei also would be smaller. As a result, the resistance to invading alpha particles would be less, enabling the alphas to make a closer approach.

Hydrogen is the lightest element and so Rutherford and Marsden set about firing alpha particles through that. The alphas were produced by a radioactive source, passed through the hydrogen gas and were detected by scintillations when they hit a zinc-sulphide screen. When the screen was more than a certain distance away, the scintillations stopped. This is because the alphas lost energy after several collisions with air molecules between the hydrogen and the screen, and came to rest at more or less the same distance from the radioactive source that had spawned them. However, occasionally a few scintillations still occurred beyond the range that the alphas had reached. By means of a magnetic field, Rutherford showed that these were caused by positively charged particles, lighter than alpha particles. He realized that the energetic alpha particles must have knocked these, lighter, positively charged particles out of the atoms in the hydrogen gas. He named these positively charged nuclei of hydrogen atoms 'H-particles'; today we call them protons.

This was a major discovery, but did not of itself establish that protons are the particles that carry the positive charge in the nuclei of all atomic elements. This major breakthrough came after Rutherford had spent three years pondering an anomaly that Marsden had found: 'H-particles' are also produced when alpha particles pass through the air.

Marsden had discovered this in 1914. He then left for New Zealand, and many other students departed to take part in World War I.

Rutherford patiently investigated the phenomenon himself, and finally established what was happening. He fired alpha particles at several light elements and found the signal of H-particles appearing. He realized that H-particles were being ejected from within the atoms of those elements; in Marsden's original experiment they had been ejected from nitrogen, which is a significant component of the atmosphere. With this new finding, Rutherford had established that H-particles are fundamental to the nuclei of all atoms; it was only then, in 1919, that he named them protons.

There is a story that illustrates the magnitude of this discovery. Rutherford's expertise was called on during the war, in which he invented ways of detecting submarines. At the climax of his experiments on protons he excused himself from a meeting of the government scientific committee saying that if what he suspected was true, it would be even more important than winning the war. Given what subsequently flowed from the atomic nucleus, and its role in the atomic bomb that ended World War II, Rutherford was hugely prescient.

By the 1920s the role of protons in carrying the nuclear charge was established, but these alone were unable to explain the relative masses of the nuclei of different elements. An alpha particle, with twice the charge of a proton, is nonetheless some four times heavier. Thus in 1920 Rutherford speculated that there is a 'neutron' – a particle of similar mass to the proton but with no electrical charge. The explanation for an alpha particle being four times heavier than a proton is thus that it contains two protons and two neutrons. By this stage, Rutherford had moved to Cambridge, succeeding Thomson as the Cavendish Professor. It was at the Cavendish Laboratory, under Rutherford's direction, that James Chadwick discovered the neutron in 1932.

Now in his fifties, Rutherford increasingly directed the research of younger colleagues rather than doing experiments of his own.

The Cavendish Laboratory contained much delicate apparatus, which could be easily disturbed or even damaged by a careless jolt. Noise itself could affect some instruments, and Rutherford's booming voice was a potential hazard. A famous photograph shows him beneath a sign saying 'Speak Softly Please'.

At this juncture, Rutherford began what has become known as 'big science': the exploitation of large instruments to explore the labyrinths of the atom. He was aware that with the proton and neutron he had identified the basic pieces of the atomic nucleus. However, the actual structure of those nuclei remained a mystery. The alpha particles, produced in natural radioactivity, had only a limited ability to penetrate the powerful electric fields surrounding the nucleus. Means were needed to increase the energy of the alphas, so that they could probe deep within the nucleus.

Thus the first 'atom smasher' was built under Rutherford's direction. With this device, John Cockroft and Ernest Walton accelerated protons – which were easier to work with than alpha particles – and fired these beams of high-energy particles at lithium. They found that a proton splits a lithium nucleus in two, converting it into two alpha particles. For the first time, artificial transmutation of the elements had been achieved.

Hitherto, radioactivity had involved the spontaneous transmutation of the elements. What had been achieved at the Cavendish Laboratory was the transmutation of an otherwise stable element into another. A new science – nuclear physics – was born. Natural radioactivity had liberated significant amounts of energy, but not in a form that promised much practical application. Rutherford's 1930s comment that anyone proposing to get useful power from the nucleus was talking 'moonshine' was, prior to artificial transmutation, correct. However, with the ability to induce transmutation, new possibilities arose. Among these was the propensity for elements such as uranium to be split, liberating neutrons that themselves induced further 'fissions' of surrounding atoms. This process has become the source of power, with both peaceful and military applications.

By the 1930s Rutherford was universally recognized as the leading experimental physicist of the day, and one of the greatest in history. Knighted in 1914, he received the Order of Merit in 1925, and by 1931 he had became Lord Rutherford of Nelson. His fame may indirectly have led to his death. In 1937 he had an umbilical hernia. British protocol at that time required that, being a member of the House of Lords, he could be operated on only by a titled surgeon. It has been suggested that the delay in finding a suitably qualified doctor contributed to his untimely death. Two years later, World War II began. One can only imagine what contributions he might have made to the Manhattan Project, where the development of the atomic bomb built upon his discoveries in nuclear physics, or to the development of radar, which was linked to his first love: electromagnetic radiation. He is buried in Westminster Abbey, near to the tomb of Isaac Newton.

Niels Bohr

Leader in quantum research

(1885–1962)

Those who are not shocked when they first
come across quantum theory cannot possibly
have understood it.

NIELS BOHR, QUOTED IN WERNER HEISENBERG,
PHYSICS AND BEYOND, 1971

Niels Bohr made major contributions to the development of quantum theory during the first third of the last century, but he was probably even more important as an inspiration and a guide to the mostly somewhat younger physicists who took the final steps. From 1921, when his Institute of Theoretical Physics was founded in Copenhagen, right up until his death, with the exception of the war years, the Institute was a world centre for research in theoretical physics. Nearly all the leading contributors to the theory spent considerable periods in Copenhagen, explaining their ideas to Bohr, working and debating with him, and assimilating much of his approach to physics.

Once the rigorous synthesis of quantum theory had been achieved by Werner Heisenberg and Erwin Schrödinger in 1925–26, Bohr developed the framework of complementarity, which claimed to 'interpret' the mathematical results, demonstrating how the apparently paradoxical aspects of quantum theory could be described rationally. Bohr also made crucial contributions to nuclear physics, but in the last period of his life he became on elder statesman of physics and a disciple of world peace.

A family of pioneering minds

Bohr was born in Copenhagen on 7 October 1885. His father, Christian, who became professor of physiology at the University of Copenhagen in 1890, was himself an excellent scientist. His most influential work concerned the effect of carbon dioxide on the release of oxygen by haemoglobin, for which he was nominated for the Nobel Prize in Medicine in 1907 and 1908. Niels was brought up in comfort, but also in an atmosphere of devotion to learning. His brother Harald (1887–1951) was an important mathematician, becoming director of the Institute for Mathematical Sciences next door to Niels's own institute. For Bohr, being a Dane, a citizen of a small country only recently humiliated in its war with Prussia and with further tribulations to come in the 20th century, was central to his life. Also significant, at least later, was the fact that he was half-Jewish.

After studies in Denmark, Bohr's international scientific career began while working with Ernest Rutherford at Manchester University from 1912. The previous year, Rutherford had discovered the nuclear atom, in which virtually all the mass was contained in an extremely small nucleus at its centre. Bohr's achievement was to link Rutherford's work with Max Planck's early quantum results. Since the foundation of the quantum theory in 1900, all the analysis had concerned radiation; the famous Bohr atom was the first application of the theory to atoms. His model of atomic structure, depicting the atom as a small, positively charged nucleus surrounded by electrons, which he published in 1913, remains in use to this day.

His work was based on quantized electron orbits around the nucleus, the angular momentum being equal to $nh/2\pi$, where n takes the values 1, 2, 3 … for the various orbits and h is Planck's constant, the most fundamental quantity in quantum theory. Bohr demonstrated the existence of discrete orbits or 'energy levels' in atoms, between which transitions could take place. That is to say, electrons could jump from one orbit to the next, absorbing or emitting radiation with a frequency determined by the energy difference between the two levels, the necessary energy being provided by a photon – a particle of light – with an appropriate wavelength. The great success of the model was to explain the wavelengths of the spectral lines of radiation absorbed or emitted by atomic hydrogen.

This work was a triumph for Bohr and justifiably gave him his Nobel Prize in Physics in 1922. Yet he was the first to recognize that it could only be a stop on the road to a full quantum theory. The physics of the day said that his atom was unstable, as the orbiting electron should lose energy and spiral in to the nucleus. A much greater break with the past was required, and this process over the next decade made much use of Bohr's correspondence principle, which stated that in the new quantum theory, the rules of so-called classical or pre-quantum physics must be replicated in sufficiently large systems. The Israeli physicist and philosopher of physics Max Jammer wrote that, 'there was rarely in the history of physics a comprehensive theory which owed as much to one principle as quantum mechanics owed to Bohr's correspondence principle'.

Arguments about quantum theory

Heisenberg, who produced the first rigorous formulation of quantum theory in 1925, was very much a protégé and follower of Bohr. It was he who deduced the famous uncertainty principle, which limited simultaneous knowledge of position and momentum; but it was Bohr who generalized the argument to an overall philosophical approach, known as complementarity, which attempted to explain the apparent contradictions of quantum theory. The complementarity principle stated that some objects have two properties that seem contradictory. Sometimes, we can switch between different views of an object to observe these opposing properties, but we cannot see both at the same time. Bohr said that we should therefore concentrate on measurement results, rather than ask why we could not discuss momentum and position simultaneously, or why light sometimes behaved as a wave and sometimes as a particle. Bohr wrote in 1936 that: 'The renunciation of the ideal of causality in atomic physics is founded logically only on our not being any longer in a position to speak of the autonomous behaviour of a physical object, due to the unavoidable interaction between the object and the measuring instruments.'

For many years Bohr's approach to these conceptual matters was unchallenged by nearly all except Schrödinger, who had come up with his own synthesis of quantum theory, independent of that of Heisenberg, and in particular, Einstein, who demanded a deeper understanding of atomic systems themselves, rather than focusing primarily on how they related to measurements. It was felt at the time that Bohr had won the celebrated Bohr–Einstein debates of the 1920s and 1930s, played out in public disputes and at occasions such as the famous Fifth Solvay Conference on quantum theory in October 1927, but more recently this belief has been questioned as theoretical physicists have challenged many of the precepts of his complementarity principle.

In the 1920s, Bohr had pioneered the technique of building up electrons in different states to provide an understanding of the atomic Periodic Table, and in the 1930s his interest switched to nuclear physics. He was responsible for the liquid drop model of the nucleus, which explained many experimental results and was important in the explanation of nuclear fission by Lise Meitner and Otto Robert Frisch in late 1938. Just like a liquid drop, the nucleus could develop a neck and split in two. Then in 1939, on a visit to the United States, Bohr worked out with John Wheeler the detailed theory of nuclear fission. They showed that the dominant isotope of uranium, U-238, does not take part in fission; it is U-235, which makes up only 0.7 per cent of natural uranium, that is fissile. Enrichment of natural uranium to increase the proportion of U-235 was central in the atomic bomb project and also in nuclear power production after the war.

A great scientific Dane

In Denmark, Bohr was by now the country's most prominent citizen; in 1931 he and his family had been elected to occupy the Residence of Honour, provided by the Carlsberg Foundation for the most influential Danish citizen in science or the arts. He had married Margrethe in 1912 and had five sons, of whom Aage became a physicist and won the Nobel Prize in 1975. They remained in Copenhagen after the German occupation in April 1940. In October the following year, Heisenberg, then working on the Nazis' nuclear programme, paid his notorious visit to Denmark, during which he antagonized the Danish physicists by assuming a German victory. In a private conversation with Bohr, he disturbed his former mentor greatly by an apparent desire to discuss the possibility of making nuclear weapons. Sixty years later, this visit became the basis of Michael Frayn's play *Copenhagen*.

In September 1943 it became known that Danish Jews were to be deported to Germany; Bohr and his wife escaped to Sweden and thence to London. In November, they travelled to the United States, joining the atomic bomb project, though Bohr was mainly concerned with the political implications. In May 1944 he met Winston Churchill, hoping to persuade him to support plans for future international control of nuclear weapons. Sadly Churchill was uninterested. Indeed, because Bohr had corresponded with Pyotr Kapitza, the Russian physicist, Churchill came close to accusing him of treason. (Bohr's message had in fact been approved by the British Secret Service.)

Bohr continued this search for peace after the war, regrettably with little success during his lifetime. For his efforts, though, in 1957 he was awarded the first Atoms for Peace award in the presence of President Eisenhower. Bohr was also influential in the planning of CERN, the European Organisation for Nuclear Research; Nordita, the Scandinavian Institute for Theoretical Atomic Physics; and Risø, a Danish centre for the industrial application of nuclear energy. He died suddenly in November 1962.

Linus Carl Pauling

Architect of structural chemistry and peace activist

(1901–1994)

*Chemistry is wonderful! I feel sorry for people who
don't know anything about chemistry. They are
missing an important source of happiness.*

LINUS PAULING, FIRST HITCHCOCK LECTURE,
UNIVERSITY OF CALIFORNIA AT BERKELEY, 1983

From the time the teenaged Linus Pauling experienced his first chemical reaction at the house of a friend in Portland, Oregon, to the final months of his life at his ranch on the Big Sur Coast of California, chemistry dominated his inner and social existence. When he proposed to the woman who would become his wife, he was honest enough to tell her that she would have to play a secondary role to his work. This passion for science proved fruitful, fostering fundamental discoveries about the nature of the chemical bond and the structures of such important biological molecules as proteins, for which he won the 1954 Nobel Prize in Chemistry. His scientific expertise was also behind his important humanistic efforts. His marshalling of evidence that fallout from above-ground nuclear tests was causing numerous birth defects and cancers played a pivotal role in his 1962 Nobel Peace Prize, presented to him on 10 October 1963, the day that the Partial Nuclear Test Ban Treaty went into effect. He thus became the only person ever to have won two unshared Nobel Prizes.

Early years of hardship and tragedy

Pauling was the first of three children and the only son of Herman W. Pauling, a pharmacist, and Lucy Isabelle Pauling, the daughter of a pharmacist. Linus spent his early years in Condon, an arid Western town in Oregon's interior where his father owned a drug-store. Some of his formative experiences involved cowboys, one of whom demonstrated the correct way to sharpen a pencil with a knife, and Native Americans, who showed him how to discover and dig for edible roots. These lessons deeply impressed the young Pauling in two ways: firstly to believe that correct techniques existed for accomplishing tasks, and secondly to understand that people with experiential knowledge were valuable sources of information. His favourite subjects in Condon's primitive elementary school were arithmetic and spelling, because they dealt with definite answers that were clearly right or wrong. Economic hardship and a fire at the store led Herman in 1909 to move his family to Portland. Not long after establishing a new drugstore, Pauling's father died suddenly of a perforated gastric ulcer at the early age of thirty-three.

Without any marketable skills, Pauling's mother (called Belle) borrowed heavily to buy a large house, hoping that roomers and boarders would provide support for herself and her children, but she often experienced money shortages and health problems, while Linus had to work at a variety of jobs, including delivering milk and newspapers. After becoming interested in chemistry, he set up a base-ment laboratory where he performed rudimentary experiments. He also took all the courses in science and mathematics open to him at Washington High School, but he left without a diploma because he failed to take the required semesters of American history (preferring mathematics courses instead). By this time, he had a well-paying job in a machine shop that manufactured freight elevators, and his mother wanted him to abandon his plans for college so that he could continue to help support the family. Fortunately, the intervention of a concerned father of one of Linus's friends convinced Belle Pauling to let her son attend Oregon Agricultural College (OAC, now called Oregon State University).

While doing exceptionally well as a chemical engineering major (the only course of study available at OAC for potential chemists), Pauling continued to work at several jobs to help provide for himself, his mother and his sisters. He was even forced to leave college for a year because of his mother's financial problems. He worked then as a paving inspector, and later, at OAC, as an instructor in quantitative analysis. This is when he began reading the papers of Gilbert Newton Lewis and Irving Langmuir on the chemical bond. He met his future wife, Ava Helen Miller, during his senior year, when he taught freshman chemistry to girls studying home economics.

After graduating from OAC in 1922, Pauling began his graduate studies at the California Institute of Technology (known as Caltech or CIT, the acronym preferred by Pauling). Besides taking a heavy load of courses, Pauling began research under the direction of Roscoe Gilkey Dickinson, an X-ray crystallographer who guided him in a structural study of the mineral molybdenite, whose sulphur atoms turned out to have a trigonal prismatic arrangement around the molybdenum atom, a surprising result that led to a published paper. After Pauling's first year at CIT, he and Ava Helen married, and she took up a role that would continue for over fifty years of marriage as primary supporter of his work in science and, later, in his peace efforts. Following his successful defence of a thesis derived from his crystal-structure papers, Pauling was awarded his doctorate in 1925.

Determining the nature of the chemical bond

In 1926 he accepted a Guggenheim Fellowship and, together with his wife, travelled to Europe, where he studied the implications of the newly discovered quantum mechanics for his work on the nature of the chemical bond – the attractive forces that hold atoms together in the form of compounds. Although he spent some time at Niels Bohr's institute in Copenhagen and with Erwin Schrödinger at the University of Zurich, he was most deeply influenced by his studies and research at Arnold Sommerfeld's Institute for Theoretical Physics in Munich. He was able to use wave mechanics (the form of quantum mechanics favoured by Sommerfeld) to predict theoretically the properties of ionic crystals.

Pauling returned to CIT in 1927 to begin a long and successful career that initially involved the X-ray study of crystal structures such as the silicate minerals, helping to make them one of the best understood branches of structural chemistry. Using his knowledge of bond distances and bond angles, he derived what came to be called his 'coordination theory', which created rules that helped crystallographers more easily to establish the correct configurations of atoms in various crystals. In 1930, due to a meeting with Herman Mark in Germany, Pauling became fascinated with electron diffraction, and, using this technique, he and his collaborators worked out the structures of many molecules existing in the gaseous and liquid states.

During the 1930s he used the interchange (or resonance) energy of two electrons in his analysis of bond hybridization (which involved the mixing of orbitals – that is to say, the space inside the atom in which a given electron can be found), which was a revolutionary idea in one of his most famous papers on the nature of the chemical bond. Pauling's grasp of quantum mechanical principles was also a major factor in his development of valence bond theory, in which he proposed that certain molecules, such as benzene, could be described as an intermediate structure formed by a combination (or hybrid) of one or more other structures with overlapping atomic orbitals. His 1939 classic *The Nature of the Chemical Bond, and the Structure of Molecules and Crystals*, based on his George Fisher Baker Lectures at Cornell University, provided a unified summary of his own experimental and theoretical studies as well as those of other structural chemists.

By the mid-1930s, Pauling's focus started to shift to biological molecules, and he and his colleagues performed magnetic studies on haemoglobin, proving that a magnet attracted haemoglobin from the venous blood but repelled it from arterial blood. Since haemoglobin is a protein molecule, his study led naturally to a more general interest in proteins, including their denaturation – the disruption of their normal structure – and their role in antibody–antigen reactions, such as when the human immune system's antibodies fight antigens attacking in the form of bacteria and viruses. His theory of the specificity of the interaction of antigen and antibody molecules, which was later shown to be flawed, was rooted in his vision of their structural complementarity.

In search of practical solutions

During World War II, Pauling's work emphasized practical problems, and he discovered an artificial substitute for blood serum that made more plasma available to wounded soldiers. He invented an oxygen detector, an instrument based on oxygen's particular magnetic properties, that found wide use in submarines and aeroplanes. He also did work on explosives, rocket propellants and inks for secret writing. Because of his battle with a serious disease, glomerular nephritis, he was unable to accept J. Robert Oppenheimer's offer to head the chemistry section of the atomic bomb project. Towards the end of the war, he learned about sickle-cell anaemia, a hereditary disease in which venous red blood cells are sickle shaped. He got the idea that this sickling must be caused by a genetic mutation in the globin part of the cell's haemoglobin. After three years of work, he and his collaborators were able to prove that such a molecular defect in haemoglobin was the disease's cause. Pauling was thus responsible for finding the first molecular disease.

In the post-war years, Pauling continued to study proteins, and in the early 1950s he published a configuration of amino acids that involved a cylindrical coil-like structure (later called the alpha helix), in which amino-acid groups were connected by hydrogen bonds. This and other protein structures that he published were extremely influential. Concomitant with this work, Pauling also became involved in efforts to educate the public about the implications of nuclear weapons. He spent more and more time in the campaign to stop their testing in the atmosphere. In January 1958 he and his wife presented to the United Nations an appeal for the end of testing that had been signed by more than 9,000 distinguished scientists. Though officials in the United States government tried to hamper his efforts by taking away his passport, they were forced to return it to him when he won the 1954 Nobel Prize in Chemistry. Throughout the remainder of the 1950s and into the 1960s, Pauling and his wife spoke all over the world on behalf of their cause, which resulted in Pauling's reception of the Nobel Peace Prize in 1963 (his wife did not share in this award because the predominantly male Nobel bureaucracy failed to nominate her).

Double Nobel laureate

Due to the negative response of CIT officials to his peace work and his Nobel Peace Prize, and because of their punitive removal of laboratory space for his work on molecular medicine, Pauling left the Institute in 1963. During the mid-1960s he was a staff member of the Center for the Study of Democratic Institutions in Santa Barbara, where his humanitarian work was encouraged and where he was able to develop a theory of the atomic nucleus (eventually rejected by most nuclear physicists). Wishing to have laboratories for his experimental research, in 1967 he became a professor of chemistry at the University of California in San Diego, where he became interested in the neglected potential of vitamin C to mitigate such problems as the common cold. In 1969 he accepted a professorship at Stanford University, and in 1970 he published his most widely read book, *Vitamin C and the Common Cold*, initiating his involvement in a controversy over megavitamin therapy that continued to the end of his life.

His views on the effectiveness of large amounts of vitamin C in the treatment of infectious diseases, cancer and other illnesses were widely rejected by the medical establishment. In 1973 Pauling and others founded the Institute for Orthomolecular Medicine (later called the Linus Pauling Institute of Science and Medicine), which had as one of its chief goals to generate laboratory and clinical evidence to verify Pauling's theories. His institute was plagued by personnel and legal problems, and Pauling himself experienced other difficulties following the death of his wife in 1981 and his discovery that he had prostate cancer in 1991. Despite these problems Pauling continued to work. In particular, he proposed arguments against the growing number of crystallographers who accepted as genuine quasicrystals with their fivefold symmetry that violated traditional rules that he defended.

What he did in the last decades of his life was what he had been doing from the time he posted tables of chemical substances and their properties above the laboratory bench in the basement of his mother's boarding house: exploring the connections between the structures and functions of molecules, not just in chemistry, but in the borders between chemistry and physics, chemistry and biology, chemistry and medicine. As an atheist and reductionist, he deeply believed that the sciences had the power to answer all the questions that humans can ask. For him, the Universe is composed solely of matter and energy, and the structures of molecules are potentially able to account for all physical, chemical, biological, even psychological phenomena. He interpreted the death of his wife and his own emotional distress after it in a rational way, as he did the cancer that spread from his prostate to his colon and finally to his liver, causing his own death in 1994. He left behind him a body of chemical knowledge unsurpassed in its richness, variety and promise for generating future discoveries.

Enrico Fermi

Creator of the atomic bomb

(1901–1954)

*This letter is a preliminary report on experiments
undertaken to ascertain whether, and in what number,
neutrons are emitted by uranium subject to neutron
bombardment, and also whether the number produced
exceeds the total number absorbed by
all processes whatever.*

H. ANDERSON, E. FERMI AND H. HANSTEIN,
PHYSICAL PEVIEW, 1939

There are a number of 20th-century physicists who were more imaginative than Enrico Fermi, one or two who were deeper thinkers, and a few who were more gifted mathematically. But Fermi, with his uncanny ability to see the essence of any physics question, was the greatest problem-solver of them all. He was also the last individual to reach the very highest echelons of the field both as a theorist and as an experimentalist.

His reaction to the explosion of the first atomic bomb illustrates these qualities. Fermi had been responsible as much as, and perhaps more than, any single individual for the physics concepts that led to the bomb's development at the Los Alamos National Laboratory in New Mexico, the research-and-design centre of the programme to develop nuclear weapons, codenamed the 'Manhattan Project'. He had also played a key role in the actual construction of the bomb itself. Many at Los Alamos considered him an oracle, to be consulted on any difficult problem of theory, experiment or numerical estimates.

Yet there is no recorded memorable phrase he uttered when the test explosion occurred on 16 July 1945. There is however a story of a different kind.

The immediate response of J. Robert Oppenheimer, the Manhattan Project's technical director, to the event is the best-known one. As he saw the blast light up the sky, a line came to him from the Hindu scripture the Bhagavad Gita, in which Vishnu announces to the Prince, 'Now I am become Death, the destroyer of worlds.' Much more prosaically, Kenneth Bainbridge, who was in charge of the test, said simply, 'Now we are all sons of bitches.' While others present at the site were trying to sort out their reactions, ranging from fear to pride of accomplishment, the always pragmatic Fermi was seen tearing up a sheet of paper into pieces, intending to find a quick and simple way of measuring the blast's impact. After some forty seconds, he tossed the pieces into the air, just as the detonation's shockwave reached his perch. He calmly observed how far the pieces had blown, then consulted a simple chart he had prepared beforehand, took out his slide rule and announced his estimate of the blast's extent. As usual, later detailed measurements showed Fermi's simple calculation was remarkably close to the correct answer. Fermi's ability to gauge the order of magnitude of any physical phenomenon was legendary and this was no exception.

Theory and experiment in Rome

Fermi was born in Rome on 29 September 1901 and grew up in an ordinary family. His father was an employee of the state-owned railways and his mother was a teacher. But his unusual aptitudes were quickly recognized and he was awarded a scholarship to attend university at Italy's elite Scuola Normale Superiore in Pisa. There he quickly outshone not only all other students, but also his teachers, Italy being at the time a relative physics backwater. This meant Fermi was essentially self-taught, developing a style of his own that emphasized stripping a problem of all but its essentials and then looking for a simple solution. This stood in sharp contrast to the then dominant German school, which relied far more extensively on mathematical analysis.

After graduation, Fermi quickly solved several important problems in theoretical physics, including an approach to statistical mechanics that incorporated the new ideas of quantum mechanics. Doing so, he attracted the attention of Orso Corbino, a much older physics professor in Rome who had been dreaming of building up a first-rate physics research institute and was looking for the right person to lead it. The politically powerful Corbino saw in Fermi the realization of his hopes and helped him to obtain a chair of theoretical physics in Rome at age twenty-six, an unheard-of precedent in Italy.

Fermi succeeded beyond Corbino's wildest expectations, attracting visitors from all over Europe to work with him and developing the talents of young Italians. On an individual level, Fermi's most famous contribution to theoretical physics, the introduction of the so-called theory of weak interaction, took place in 1934. It had been known for a number of years that nuclear decays in which an electron is emitted seemed to violate the conservation of energy, a great puzzle since this principle was regarded as the bedrock of physics. Niels Bohr had suggested that the conservation might not be quite so absolute. Conversely, Wolfgang Pauli held that it was and that the missing energy was carried away by a particle that escaped detection. But how? In 1934 Fermi showed a way in which it could occur. Calling Pauli's particle a *neutrino*, he postulated the existence of a fundamental new kind of interaction, one that allowed a neutron to decay into a proton, an electron and a neutrino. He showed the form the interaction might take, estimated its magnitude and explored its consequences. At a time when the only two known forces were gravity and electromagnetism, this was a revolutionary concept; it has since come to be seen as a milestone in physics.

While continuing to work as a theorist, Fermi also helped to form a close-knit group of experimentalists, most of whom later went on to have distinguished careers of their own. The original cluster, consisting of Edoardo Amaldi, Bruno Pontecorvo, Franco Rasetti and Emilio Segrè, participated in what remains probably Fermi's most important experimental endeavour. Up until the early 1930s, using a technique pioneered by Ernest Rutherford, the process of scattering from nuclear targets had chiefly been achieved by firing beams of alpha particles (that is, helium nuclei) obtained from radioactive decays at the target. But the discovery in Cambridge by James Chadwick in 1932 of the neutron meant a new projectile was available. It was admittedly harder to focus a neutron beam, but their electrical neutrality meant neutrons would not be repelled when bombarding a nucleus and therefore were more likely to reach the target. Moreover, Fermi now had an important insight. It had been expected that the probability of a nuclear transformation would increase as the bombarding neutron's energy rose, but he realized that the opposite was true. The slower the incoming neutrons, the more time they spent traversing the target nuclei and the more likely they were to interact. This new system led to a very important set of discoveries, including that of nuclear fission in 1938, some made by his group and some by others. Fermi was awarded the 1938 Nobel Prize in Physics for this work and left directly from Sweden for the United States, wisely choosing to emigrate since his wife was Jewish, and Italy under Mussolini's rule had just passed a set of draconian racial laws. Fermi's departure marked the end of an era in Italian physics, but the subject had now entered the country's mainstream and would remain there despite the loss of its greatest practitioner.

The world's first nuclear reactor

Fermi's research with neutrons continued in his new homeland, though increasingly directed towards military purposes. He led the construction of the first nuclear reactor, built at the University of Chicago in 1942, and was at the controls when criticality – the point at which a reactor can self-sustain a fission chain reaction – was reached. Afterwards he transferred his operations to Los Alamos, but at the war's end he returned to the University of Chicago and became a leader in the emerging field of high-energy physics, while also branching out into other areas such as astrophysics. Maintaining his habit of working both as a theorist and experimentalist, he continued his legendary career, attracting many of the most talented young American physicists to Chicago.

In 1954, while at the very height of his powers, Fermi was diagnosed as suffering from stomach cancer and died soon after an exploratory operation revealed the cancer had metastasized. He was universally grieved: the United States' largest high-energy facility was named the Fermi National Laboratory. More to the point, all elementary particles with half-integral spin (an intrinsic property), a list that includes the neutron, the proton, the electron and the neutrino, are now known as fermions.

Hideki Yukawa

Japan's first Nobel laureate

(1907–1981)

The results of physics are inevitably connected with the problems of humanity through their application to human society; one cannot be blind to this connection.

HIDEKI YUKAWA, IN AN OPENING ADDRESS AT THE
FIRST KYOTO CONFERENCE OF SCIENTISTS, 1962

Until the 20th century, the contribution of Japanese scientists to physics was limited. All that was to change with Hideki Yukawa. He was born in Tokyo, a year after his friend and fellow Nobel Prize-winning physicist Sin-itiro Tomonaga. Hideki was one of seven children and grew up in Kyoto, where his father Takujo Ogawa was a professor of geography. As was the custom, he was known as Hideki Ogawa until he married a Japanese dancer by the name of Sumi Yukawa and took on her family name.

Hideki was a bright child but, in common with many of the greatest mathematicians and physicists, was uninterested in worldly affairs and social interactions. In fact, by his own admission he was clumsy and ill at ease in his relationships with other people, preferring the abstract world of mathematics. While at high school he was inspired by Albert Einstein's visit to Japan in 1922, and soon after by the discovery of a book, in German, by one of the pioneers of quantum theory, Max Planck. In 1926, he began studying physics at his local university in Kyoto. This was where he first developed his lifelong friendship with Tomonaga.

After graduation, then marriage, Yukawa was appointed as a lecturer at Kyoto in 1933. He now began to think in earnest about the attractive force that binds atomic nuclei together. The second constituent of the nucleus, the neutron, had just been discovered and, since it was electrically neutral, it could not bind with the positively charged protons via the electromagnetic force that held the electrons in orbit outside the nucleus. It was quickly realized that a new force had to be at work within the tiny confines of the nucleus, but no one really understood its origin or properties. In theoretical physics, quantum mechanics had been developed by several European physicists just a few years older than Yukawa, such as Heisenberg, Pauli, Dirac and Fermi. Dirac in particular had proposed what is called a quantum field theory in which forces between particles such as electrons are mediated by fields that can be described at the quantum level by what are called exchange particles. Yukawa began work on a quantum field theory that would describe the glue that held protons and neutrons (collectively called nucleons) together within nuclei.

Theorizing the heart of the nucleus

Shortly before his big breakthrough in 1935, Yukawa had moved to Osaka University. It was there that he published his famous paper, 'On the interaction of elementary particles', in which he proposed that the exchange particle between nucleons would have a mass somewhere between that of an electron and a nucleon, which was two thousand times heavier. In his theory, Yukawa suggested that the way two nucleons bind together is by utilizing what is known as Heisenberg's uncertainty principle, in which a tiny amount of energy can be 'borrowed' by a subatomic particle from its surroundings for a very short amount of time. The more energy that is borrowed, the less time the particle can keep it before having to 'repay' it. Einstein's famous equation, $E = mc^2$, tells us that since mass and energy are interchangeable, this borrowed energy can be used to create a particle of a certain mass. Yukawa proposed the creation of such a particle, now called a meson, inside nuclei. This particle, he suggested, was responsible for the attractive force between protons and neutrons. Yukawa's calculations predicted that a meson is created by one nucleon, which borrows enough energy from its surroundings to create it, then the meson jumps across to a nearby nucleon whereby it vanishes again. During its brief existence, it is interpreted as being exchanged by the two nucleons, giving rise to an attractive force that pulls them together.

The following year, experimentalists thought they had discovered just such a particle created in cosmic rays, but it later turned out that this 'mu meson' was in fact a relative of the electron and had no role to play within the nucleus. The first true meson to be discovered (called a pion, short for 'pi meson') had to wait until 1947; it was found by Cecil Powell, César Lattes and Giuseppe Occhialini at the University of Bristol, in England. Two years later, Yukawa received the Nobel Prize in Physics for his theory. By this time, he was working in the United States, first at the Institute for Advanced Study in Princeton and then at Columbia University in New York. In 1953, he returned to Japan, where he headed up a new research institute in Kyoto, which to this day bears his name. He spent the remaining years of his life working in theoretical particle physics.

LIFE

The modern life sciences began with Robert Hooke, the curator of experiments at the Royal Society in London, who invented a microscope capable of magnifying from 50 to 100 times. In 1664, Hooke published *Micrographia*, an amazing folio of drawings of magnified needles, soot, flies and fleas, linen, mould, cork, feathers and more, including a plant cell – a term coined by Hooke because its shape reminded him of the cell of a monk. Samuel Pepys bought the book immediately and noted in his famous diary that *Micrographia* was 'the most ingenious book I ever read in my life'. Soon, the Dutch microscopist Antonie Van Leeuwenhoek launched the science of microbiology with his pioneering observations – published in the Royal Society's *Philosophical Transactions* from the 1670s – of protozoa, bacteria, spermatozoa, muscle fibres, blood flow in capillaries and the fine structure of plants.

In the 18th century, the mechanism of plant respiration was established by another Dutchman, Jan IngenHousz, who measured the gaseous production of vegetables. He proved that the green pigment in plants (later named chlorophyll) had the capacity to absorb energy from sunlight and use it to synthesize carbohydrates from carbon dioxide and water, in the process producing the oxygen breathed by animals. He also showed that at night, in the absence of sunlight, plants emit carbon dioxide.

Crucial though this discovery of photosynthesis was, the most pressing requirement in biology was a widely agreed classification system for the natural world. John Ray made a start with his three-volume history of plants, completed in 1704, which contained the first biological definition of a species. But it was only in 1751 that

Carl Linnaeus produced his binomial system, still used today. In the Linnaean system, the first part of the Latinized scientific name is the generic name or genus, and the second part is the specific name or species; both are italicized. For example, in the name of the common frog, *Rana temporaria*, the genus is *Rana* and the species is *temporaria*. With the creeping buttercup, *Ranunculus repens*, and the meadow buttercup, *Ranunculus acris*, there are two species, *repens* and *acris*, that belong to the same genus *Ranunculus*. The discoverer's name may be added to the binomial name in an abbreviated form in roman type – for instance, the scientific name of the common daisy is *Bellis perennis* L. (where L. stands for Linnaeus). Thus, whereas the common name of an animal or plant puts the species first and the genus second, the scientific name reverses this order.

Linnaeus, however, was a creationist, who believed that God had created all the species in precisely their present form – despite his discovery of hybrid plants, formed by the crossing of species. Not until a century after Linnaeus's system did a scientific understanding of hybridity begin to develop. In 1859, Charles Darwin published his *On the Origin of Species by Means of Natural Selection*, which rejected creationism and put forward his principle of natural selection to explain the transmutation of species. Yet, Darwin was unable to propose a satisfactory biological mechanism for natural selection that would account for the inheritance of characteristics by plants and animals from one generation to the next: his theory of 'pangenesis' and 'gemmules' was wrong.

The answer came from a contemporary of Darwin's, the Austrian monk Gregor Mendel, of whom Darwin was unaware. During the 1860s Mendel made extensive breeding experiments on garden peas. By observing the characteristics of successive generations and hybrids, Mendel defined a theory of inheritance based on elements within the pea's reproductive cells, later to be known as genes, which became the foundation of genetics. In 1953, the genetic mechanism of replication and inheritance was revealed by the physicist Francis Crick and the biologist James Watson, who discovered the 'double helix' structure of deoxyribonucleic acid (otherwise known as DNA), thereby creating the discipline of molecular biology. This subsequently transformed the life sciences, including the study of the nervous system, which had been pioneered in the 19th and early 20th century by, among others, the physiologist Jan Purkinje and the neuroscientist Santiago Ramón y Cajal. Although neuroscience has yet to account for human consciousness, as Crick hoped it would, it is a promising field for major scientific advances.

Carl Linnaeus

Botanist who named the natural world

(1707–1778)

*I saw at an infinite distance omniscient and almighty
God, where He walked His way; and I marvelled. I traced
His footprints in the fields of Nature and I noticed in every
thing, even that which was hardly visible, an endless
wisdom and power, an inscrutable perfection.*

CARL LINNAEUS, INTRODUCTION TO *SYSTEMA NATURAE*,
12TH EDITION, 1766

In many ways Carl Linnaeus has become a symbol for Swedish science in the 18th century. He was not only one of the pioneers in the new scientific development of Sweden, he also became internationally well known and as a result drew attention to the country's scientists. Though he was ennobled in later life, he came from humble roots: born on 23 May 1707 in rural Småland in southern Sweden into a family of peasants and priests. Thanks to his father, Nils, a vicar in a small village parish and an amateur botanist, the young Carl soon became interested in flowers and plants. After several years at the gymnasium school, he moved in 1727 to the University of Lund, but after only one year he decided to enrol at the University of Uppsala. He wished to train as a doctor, but it was above all botany, which played an important role in medicine, that most fascinated him.

Interpreting natural forms

He quickly saw that the discipline had entered into a confused and complicated phase. As a consequence of travel to foreign lands, knowledge of new plants was dramatically increasing. But whether one followed Aristotle or recent systematists such as Andreas Caesalpinus, Caspar Bauhin or Joseph Pitton de Tournefort, there arose problems. Some naturalists divided plants according to colour, others according to size, still others according to the corolla and fruit. Linnaeus, too, got the idea of constructing his own system. He had learnt from the German Rudolf Jacob Camerarius and the Frenchman Sébastien Vaillant that plants were endowed with sexuality; their stamens with pollen could be regarded as the equivalent of male genitals, and the pistils with their ovaries, the equivalent of female sexual organs. By the end of 1729, Linnaeus had penned a little essay in Swedish with the Latin title 'Praeludia sponsaliorum plantarum' (Foreplay to the wedding of plants). Later, he would go even further on the basis of new observations.

The professor of botany at Uppsala at that time was old, and Linnaeus was able to take over the teaching of students and to lead them on excursions. In the summer of 1732, he made a trip to Lapland and wrote an account of the journey, which would later be printed as *Iter lapponicum*. It was first published in English in 1811, and in Swedish in 1889.

To become a physician in Sweden it was necessary to undertake a study tour abroad and to obtain a doctoral degree at a foreign university. In April 1735, Linnaeus set out on his own journey and, passing through Hamburg and Amsterdam, made his way to the small university town of Harderwijk in the Netherlands, where he qualified as a doctor with a thesis on fever. He then set off for Leiden to meet the most eminent medical oracle of the time, Hermann Boerhaave, as well as the oustanding botanist Johann Friedrich Gronovius. Linnaeus had the good fortune to find employment with George Clifford, a rich director of the Dutch East India Company, on his estate Hartekamp, located between Leiden and Haarlem. Here he would stay for two years in charge of Clifford's garden, library and herbarium.

A new system of classification including *Homo sapiens*

At Hartekamp, Linnaeus's research was hectic. He worked on several manuscripts that he had brought with him from Sweden and that he now had printed. His obvious talent caused patrons and supporters to compete with each other, and he was supported both economically and morally by Gronovius as well as Clifford. The most important of his published texts from this period was *Systema naturae*, which appeared in late 1735. It contained in large folio format a division of nature's three kingdoms in the form of tables: the mineral world (*regnum lapideum*), the plant world (*regnum vegetabile*) and the animal kingdom (*regnum animale*). Highest among the four-limbed animals, Linnaeus placed the human being, *Homo sapiens* – a term he himself coined. This proposal was of course daring at the time and could cause unease. Linnaeus retorted in his defence that man was part of the Creation and clearly did not belong to the mineral or vegetable kingdom but to the animal kingdom. He was also the first to divide humankind into races or varieties, and in doing so he founded physical anthropology, a field that has never been free from complication. He estimated that there were five races: the American, the European, the Asian, the African and a fifth, 'monstrous' mixed group, to which he assigned the Hottentots, among others. It will scarcely come as a surprise, given the period, that white Europeans came first among the races, and black Africans, last.

But the most celebrated of the realms set out in *Systema naturae* is the plant world. More specifically, it described the sexual system of plants, which has become known as the system of Linnaeus. In the book, he proceeded from his earlier conviction that plants were sexual creatures and so based his system on their reproductive organs, the stamens and pistils. By counting the stamens and noting their arrangement, he divided plants into twenty-four groups or classes; and by counting the pins of the pistils, he divided them into subgroups or orders. The divisions ranged from classes, through orders

and families, down to species. The ten first classes included plants
with one to ten stamens; the subsequent thirteen classes contained
plants with different arrangements of the stamens (such as two long
and two short stamens); while the twenty-fourth class contained
flowerless plants, *cryptogamia*. Just as in his youthful essay on the
marriage of flowers, in *Systema naturae*, Linnaeus gives a vivid and
poetic description of the sexual life of plants. Of the first class he
writes that the flower has one man in marriage, whereas in the
eighth class there are eight men in the same bridal chamber with
one woman, and in the fourteenth class, four men, two of whom
are tall and two short, are with the bride. Some readers were upset
by this outspokenness. Nonetheless, his sexual system was quickly
accepted and became of great importance. It can be said that Linnaeus
gave botanists a common language; it became easier for them to
communicate with each other. During Linnaeus's lifetime, *Systema
naturae* was published in a large number of new editions. The last
from his own hand was the twelfth edition in 1766–68; by then
the original eleven pages had swelled to three volumes. Linnaeus's
sexual system is still in use in elementary handbooks, although no
longer in more advanced scientific literature. Linnaeus himself never
intended his system to represent natural groups, but only to be used
for identification.

His productivity during the three years in Holland was almost
incredible. In addition to *Systema naturae* and some shorter pieces,
he published eight large works, several of which formed an extension
or application of the sexual system. He fulfilled his official duties
in the form of his impressive description of the flowers in Clifford's
garden, *Hortus Cliffortianus*. Later came his own original work. In
Fundamenta botanica, he explained the actual method underlying his
system, how plants should be placed in species, orders and classes, as
well as describing them. In *Critica botanica*, he provided the rules for
the naming of species. *Genera plantarum* catalogued and described

all of the plant families he was then concerned with, organized into classes and orders, and gave a historical account of all the botanic systems, from that of Caesalpinus to his own. Remarkably, he also found time to visit England in the summer of 1736, at Clifford's expense. In London he met Sir Hans Sloane, Isaac Newton's successor as president of the Royal Society, and in Oxford, the German botanist Johann Jakob Dillenius.

Linnaeus returned to Sweden in June 1738, never to leave the country again. He settled down as a practising physician in Stockholm. In 1741, he became a professor at the University of Uppsala and was appointed physician at the royal court. In 1758 he purchased the farm of Hammarby, which lies outside Uppsala and is still in use as a museum, and in 1762 he was ennobled under the name von Linné.

Naming the natural world

If the description of the sexual system in *Systema naturae* was Linnaeus's first great achievement, the second was his *Species plantarum* of 1751, dealing with the species of plants, which is accepted as the starting point of modern botanical nomenclature. This covered all the recognized species then known to him, about 8,000 in all. Equally important was how they were to be registered. Here Linnaeus proposed the use of his binomial nomenclature, that is, a two-name system with a forename and a surname for all species. Previously species had been given a family name followed by a long description. Now, with only two words, Linnaeus could clearly identify a plant. The first word described the family (*genus*); the second, the species – for example, *Sinapis arvencis* for wild mustard. The names were not allotted by chance. Family names were commonly attached to the name of some famous botanist. Linnaeus's colleague Gronovius named an undistinguished little plant, common in Sweden, after Linnaeus and called it *Linnea borealis*, in which the element 'borealis' means 'northern'. Linnaeus regarded this as a good choice, since he, like the little flower, was 'modest and unpretentious' – although perhaps not all of his colleagues would have agreed.

His third major contribution was perhaps his manner of describing plants. He defined the different species much more clearly and introduced a unified terminology for the parts of plants that were important for identifying them. He had a sharp eye for small details; his descriptions were always definite, concise and clear and with a few words conveyed the essentials. Linnaeus carried his urge to classify to extremes. Some thought that the structure of nature as a result was reduced to an abstract model, a sort of theorizing. Particularly critical were his French colleagues, the Comte de Buffon and Michel Adanson. Buffon claimed that a scientific classification was in fact impossible, since both the animal kingdom and the world of plants contained forms that glide into each other and cannot be distinguished by means of sharp boundaries.

Science grounded in faith

Linnaeus's view of nature was based on his religious beliefs. As he saw it, there was a firm structure in the Universe. A species was constant and unchangeable and the number of species was the same as it had been at the Creation. God had created all the species in precisely the form they now have. Out of this creative act everything had arisen in a natural way from an egg or from a seed. Linnaeus frequently cited the phrase '*omne vivum ex ovo*', which means 'all that lives comes from the egg'. This suggests that he did not accept the idea of a primordial creation, or the popular belief that certain insects had come into being from the corpse of an animal. But after a while Linnaeus was compelled to question his notion of the constancy of the species. He discovered a new plant, *Peloria*, which he first saw as a deformed specimen of *Linaria vulgaris*, but then understood to be a hybrid, that is, a cross between species. Confused, he described the plant in his 1744 dissertation *Peloria* and attempted to explain why it did not fit in with his classification system. Forced now to accept hybrids as a fact, he thought instead that all species within a family were derived from one and the same mother-form.

Despite the obvious scholastic strains in his system, Linnaeus was never an armchair thinker who abandoned nature. No one could describe natural beauty as he did; no one celebrated God's nearness as he did. In his classical texts he included passages on 'the remarkableness of insects', on 'sweetness in Nature' and on 'wonder before Nature'. God was everywhere present in his Creation, and the scientist had the task of demonstrating this. Thus, Linnaeus subscribed to the ancient Aristotelian idea of the 'chain of nature', *catena naturae*, according to which everything in the Creation could be ordered in a hierarchy, from the angels at the top, through human beings, animals and plants, down to lifeless matter. In the Creation there were no gaps: 'Nature makes no leaps', Linnaeus declared.

The Divine Order could also be described in another way, as a state of equilibrium. It was this that Linnaeus took up in a work entitled *Oeconomia naturae* (Nature's economy), published in 1749. All living creatures depend on each other for their survival. The defeat and death of one individual always leads to the gain of another. What happens is 'a war of every man against every man'. So it was in society, Linnaeus believed, for war occurs most often in well-populated areas and holds down the growth of the population. Order and balance could be observed in nature in many ways. Linnaeus never tired of repeating his ideas about how plants and animals spread geographically. Different life forms required different sorts of environment. God had established on Earth different climates and milieus so that all creatures should be satisfied. It was in this way that the principle that the Creation contained all forms of life with no gaps was fulfilled. Linnaeus, therefore, by virtue of his thought on different plants and their different demands for special conditions in order to survive, can be seen as a forerunner of what is now called ecological thinking.

Yet another side of Linnaeus was his interest in popular beliefs and his mysticism. Despite living in the middle of the 18th century when enlightened belief in reason had spread throughout Europe, Linnaeus could be strangely old-fashioned. He accepted, for example, the popular belief that swallows spent the winter at the bottom of the sea, but he never made an experiment by putting a swallow under water. He was sympathetic to the notions of numerology and to the idea that a human being passes through twelve periods, each of seven years: at seven the child loses its milk teeth, at fourteen comes puberty, and so on. Linnaeus also wrote a treatise on various creatures in the animal kingdom. This group included not only chimpanzees and orang-utans, but also certain animal-men, cavemen and men with tails. Of course, Linnaeus himself had never seen these fantasy creatures, only read about them or heard tell of them. Yet, this did not prevent him from publishing images of them. At the same time as he pretended to be a rational scientist placing man in the animal kingdom, Linnaeus naively believed in superstitious folk tales.

Linnaeus had a special relationship with his students. He looked after them, spoke lovingly of them, and sent them out into the world in search of rare plants. He called them his 'apostles'. He was driven by a desire to learn as much as possible about the created world, most of all every plant and every animal. He sent his 'apostles' all over the globe from Iceland in the north to Australia in the south, from Japan in the east to America in the west. They brought their finds home to the Master in Uppsala; they wrote letters and reports to him; and published their findings in learned journals or books. Linnaeus never in his mature years himself travelled in Europe, but through his apostles he continued to investigate nature in every corner of the world. During the last four years of his life, Linnaeus could not do any scientific work, after suffering two strokes that left him semi-disabled. He died in his home in Uppsala, aged seventy-one, a respectable age for his time.

Jan IngenHousz

Physiologist and discoverer of photosynthesis

(1730–1799)

*There are indeed very few new experiments or discoveries,
which are capable of being turned to any immediate
advantage except perhaps that of surprise or admiration,
and in the discoverer a kind of delight, unknown to any
but true rational minds, mixed with a kind satisfaction
bordering on a degree of irresistible pride inseparable
from the consciousness of having enlarged the
boundaries of human knowledge.*

LOOSE NOTE BY JAN INGENHOUSZ SCRIBBLED ON
A PIECE OF PAPER, DATE UNKNOWN

When the word 'photosynthesis' was used for the first time in 1893,
more than 100 years had passed since the fundamental characteristics of this biochemical process had been described. In the summer
of 1779, the Dutch doctor Jan IngenHousz performed an elaborate
series of some 500 experiments in a country house near London and
described his findings in *Experiments upon Vegetables*. By the time this
discovery was finally named, everyone had forgotten the name of its
discoverer. Today, it is reappearing from the mists of time. Rightly
so, as IngenHousz was not only a gifted experimenter, a talented
medical doctor and a prolific researcher in chemistry and physics,
but also a critical investigator and a polyglot traveller.

Jan IngenHousz was born in Breda, close to what is now the
border between Belgium and the Netherlands. Being Catholic, he
was barred from the Protestant universities in his native country

and had to go across the border to Louvain to study medicine. After graduating, he specialized in Paris, Leiden and Edinburgh, satisfying his curiosity in courses on gynaecology, physiology, agriculture, chemistry and pharmacology. After his father's death, he left a flourishing medical practice in his hometown in order to accept the standing invitation of Sir John Pringle to come to England. Pringle knew the young and talented IngenHousz from the time he had spent on the continent with the British army. Pringle was by now the celebrated author of *Diseases of the Army* and royal physician to King George III. He introduced his protégé to the scientific and political elite of London. Benjamin Franklin was one of them; he and IngenHousz would remain lifelong friends and scientific partners.

In 1766, IngenHousz became engaged in the campaign to inoculate the population against the killer epidemic smallpox, using live smallpox virus. This was one of the first effective prevention campaigns in the history of medicine. While it provoked a lot of resistance, mainly in religious circles, it also enjoyed the interest of the enlightened elite across Europe. IngenHousz was invited by Empress Maria-Theresa of Austria to come and treat her family in Vienna. The treatment was successful and IngenHousz was appointed royal physician, with an annual stipend for the rest of his life. Now rich and independent, he travelled and met intellectuals, politicians and natural philosophers all over Europe, living in London, Paris and Vienna, and spending long periods at Bowood House in Calne, the Wiltshire mansion of the Earl of Shelburne, where Joseph Priestley had a laboratory and discovered oxygen in 1774. IngenHousz was elected a fellow of the Royal Society in 1779, but he turned down most invitations from other societies and even refused an imperial request to become head of all universities and libraries in Austria. A side effect of his relentless quest for trustworthy knowledge led to his fierce critique of the quack Anton Mesmer, which drove the inventor of 'animal magnetism' from Vienna.

The secret life of plants

His pragmatic mind led IngenHousz to optimize the design of the eudiometer. This instrument to (literally speaking) 'measure the good-ness of the air' had been conceived by Priestley and developed by Abbé Fontana, one of IngenHousz's peers. The device was not only at the roots of the environmental sciences, but also enabled him to study the gaseous production of plants. As the subtitle of IngenHousz's 1779 book makes clear, he understood the essence of this unsuspected vegetable mechanism: *discovering the great power of purifying the common air in the sun-shine, and of injuring it in the shade and at night.* To express this in modern terminology: plants build carbohydrates from hydrogen and carbon, which they draw from water and carbon dioxide. They power this process with the energy from the Sun, captured by the green molecular complex chlorophyll. The waste product of this process is oxygen, the life-giving gas for all animals, including ourselves. IngenHousz eliminated in a truly empirical way all irrelevant variables and described how only the green parts of plants 'purify' the air, producing 'dephlogisticated air' (oxygen). By comparing the 'airs' from plants in sunshine, in the dark, or near the stove, he demonstrated that they 'improve the goodness of the air' with the help of the light of the Sun and not by its heat. He also showed that they 'injure' the air by producing 'fixed air' (carbon dioxide), just like any other breathing organism. He could best observe this in the dark, when the production of carbon dioxide was not outpaced by the oxygen production.

The publication of his findings in the autumn 1779 was the begin-ning of a lifelong dispute with Priestley, Jean Senebier and Willem Van Barneveld, who claimed to have been the first to discover the process and contested IngenHousz's claim. However, fully approved by his friend Franklin, IngenHousz preferred to get on with his research and its applications, rather than continuing with endless polemics. He went on to underline plants' crucial role in the regulation of the Earth's environment. Light was the secret behind the 'vegetable economy' and was driving the world ecosystem. He experimented with methods to improve the growth of plants, directly relevant to acute needs in agriculture, and became an honorary member of the Board of Agriculture when it was established in London in 1793. By 1789, some ten years after his landmark description, he had rephrased the interaction of plants, sunlight and atmosphere in the new chemical terminology of Lavoisier, with oxygen, carbon dioxide and hydrogen as the ingredients for the photosynthetic reaction.

IngenHousz's photosynthesis research was only one of his many endeavours. In 1785 he described the use of the cover slip for the microscope and the random motion of particles in a solution, now known as 'Brownian motion'. He further worked and published on electrical conductors, machines and pistols, on lightning rods and cannon powder, on magnetism, the properties of metals, and lamps burning inflammable gas. The medical applications of new findings also interested him greatly. He described the first use of electroshock to treat psychiatric disorders, and designed an apparatus to treat people with various ailments through inhalation of oxygen, contributing to the invention of 'pneumatic medicine' by Thomas Beddoes.

Fleeing Paris to escape the French Revolution in July 1789, he returned to London. The way home to Vienna and his wife would remain blocked by the political turmoil for the rest of his life. During his last years he struggled to finish the many experiments and articles on his drawing board. In a last bid for evidence-based medicine, he corresponded with Edward Jenner about the new vaccination method the latter was promoting, which used cowpox instead of smallpox. Jenner's new technique seemed promising, but IngenHousz feared that it could be dangerous to use without sufficient evidence of its safety. He died at Bowood House in 1799; nobody knows where in the church of Calne his body is buried.

Charles Darwin
The theory of evolution by natural selection
(1809–1882)

*Nineteen years ago it occurred to me that – whilst
otherwise employed on Natural History – I might perhaps
do good if I noted any sort of facts bearing on the
question of the origin of species.*

CHARLES DARWIN, IN A LETTER TO
ASA GRAY, 20 JULY 1857

No list of famous scientists would be complete without Charles Darwin, the man who posited the theory of evolution by means of a mechanism he called 'natural selection'. Yet Darwin himself hardly ever used the word 'scientist', and never about himself. Celebrated for major advances in geology, he rarely called himself a geologist; responsible for much of our understanding of plant physiology, he denied being a botanist. Instead, Darwin thought of himself in the broadest terms as a 'naturalist', and it is the breadth of his enquiries, crossing boundaries between disciplines, that characterizes his science and lies at the heart of his success.

There were many strands to Darwin's scientific training: the son of a medical doctor, he was the grandson on one side of the famous inventor and philosopher Erasmus Darwin, and on the other of the technologist Josiah Wedgwood I, founder of Wedgwood Pottery. Growing up in Shropshire, he and his brother had a passion for chemistry, money for chemicals and space enough for what they called a laboratory. It was the only laboratory Darwin ever had – his science began and remained remarkably domestic.

As a younger son, Darwin needed a profession. There were few respectable possibilities. The most obvious, to become a doctor like his father, led him to study medicine at Edinburgh for a time, but he left without graduating, and moved to Cambridge in 1828 to study for a general degree that would allow him to go into the Church – a common career path for someone of his social class. At Edinburgh, Charles had already demonstrated a serious interest in natural history: he disliked and ignored his anatomy lessons, but after hours spent on the seashore with the zoologist Robert Grant, he wrote his first known scientific work, a paper on the seaweed-like bryozoan, *Flustra*. At Cambridge, popular and easy-going, he rode, and went to concerts and parties. But he was also an avid collector of beetles, and joined the social circle of two leading academics, both of whom played an important part in his scientific education: Adam Sedgwick, professor of geology, and John Henslow, professor of mineralogy and of botany.

From amateur naturalist to respected scientist

Returning home from a post-graduation field trip with Sedgwick, Darwin found a letter from Henslow offering him a place on an Admiralty surveying voyage to South America. He would join HMS *Beagle* as a freelance naturalist and companion to its captain, Robert FitzRoy. Recommended as much for his social standing and his even temperament as for his fledgling scientific skills, he was not, said Henslow, 'a finished naturalist'.

The *Beagle* set off on 27 December 1831. Darwin later remembered this voyage round the world, originally planned for two years and stretching to five, as life-changing, and it was. He left an untried twenty-two-year-old novice with no long-term plans, and returned an admired member of the scientific establishment.

Over the course of the voyage Darwin learned the practical skills of a scientist: observation, collection, preservation, meticulous record-keeping, classification, the use of a microscope. He shipped back specimens of plants, birds, insects, fossils and all kinds of marine creatures, by the crate and barrel load. These collections were important but not unique: there were more collectors in South America, Darwin joked, 'than Carpenters or Shoemakers or any other honest trade'. What differentiated Darwin was his awareness of the context of what he collected, coupled with the comparisons he was able to draw across vast geographical areas and, as he began to collect fossils, across vast chronological ones also.

Travelling hundreds of miles on horseback over the Andes, Darwin saw evidence of almost unimaginable change in the landscape. Fossilized trees once submerged in seawater now stood at the highest pass in the mountains. Back with the ship following an earthquake, FitzRoy showed him evidence from the survey data of a small but permanent change in the relative height of land and sea. Charles Lyell's newly published theory that the present-day landscape was the result not of single cataclysmic events, but of the gradual operation over millennia of known causes, fitted perfectly with what Darwin observed. Literally piecing together his observations across the continent, he assembled long geological cross-sections on bits of paper haphazardly glued together, and proposed an ambitious theory of the Earth's crust as giant blocks that rose and fell and tilted on the molten core beneath.

Although Darwin claimed that he 'gathered blindly all sorts of facts' and then drew general conclusions from them, his methods were in practice more complex and subtle. 'Great quantities of facts', in particular when marshalled to support his published arguments, were at the heart of his work, but he was never afraid to build ambitious hypotheses at an early stage and then look for the data against which to test them. A surprising number of the ideas Darwin only published decades later – including the germ of his species theory – are already postulated in the notebooks that he started to keep soon after his return to England in October 1836. By this time, he was already something of a celebrity within scientific circles, thanks to Henslow having circulated a pamphlet of his geological letters. He now found himself being fêted by the science elite and offered several writing and research opportunities.

Applying his hypothesis of subsidence and uplift, Darwin proposed solutions to two hotly debated questions in geology, in one case with spectacular success and in the other leading to what he called 'one long gigantic blunder'. Being on a naval vessel, Darwin was acutely aware of the dangers of oceanic coral reefs. Their presence seemed inexplicable, given the inability of coral polyps to live in water more than sixty metres deep, and Lyell had recently argued that they must grow on the cones of underwater volcanoes. Armed with his theory of large-scale subsidence, however, Darwin suggested that they originated in shallow water around islands, keeping pace, one generation on top of another, as the islands themselves gradually became submerged. This elegant solution ensured Darwin's place among serious scientists. Back in Britain, he was elected secretary to the Geological Society in March 1838 and toyed briefly with the prospect of a university career.

Darwin now turned his attention to the so-called 'parallel roads' of Glen Roy in the highlands of Scotland – a series of striking terraces running round a system of valleys off the Great Glen. Geologists argued that the valleys, once thought to be man-made, formerly held lakes whose waters had cut the 'roads' as they rose and fell. But there was no trace of the series of huge dams that this implied. Darwin suggested, quite neatly, that it was rather seawater, draining away as the land mass slowly rose, that was responsible. There were problems with this too, however, in particular the lack of marine fossils, and soon after it was published, Darwin's theory was challenged by Louis Agassiz, who proposed that the elusive dams had been formed by glacial ice. It was a lesson to Darwin to temper the grand scope of his theorizing with greater caution in what he made public.

Shortly after his one field trip to Glen Roy, in January 1839 Darwin was elected a fellow of the Royal Society. Five days later he married his cousin Emma Wedgwood, and they soon began a family. Ever the observer, he opened a notebook to record every aspect of the behaviour of his infant son, William, who was born in December of that year. His notes on William were first used in his book *The Expression of the Emotions in Man and Animals*, published in 1872, and then again in 1877 in 'A biographical sketch of an infant', a paper in the newly established psychological journal *Mind*.

The evolution of evolutionism

In 1842 Darwin and his growing family moved to the village of Downe in Kent, nicely calculated to be close enough to London to maintain professional and personal ties, but far enough to deter casual visitors. It remained his home – and his workplace – for the rest of his life, his wife and children providing the loving and stable environment he needed. By this time, Darwin was an established popular author, with his account of the *Beagle* voyage a bestseller. He was also chronically ill. Nevertheless he began a friendship that was personally and professionally crucial. A young botanist and explorer, Joseph Hooker, later director of the Royal Botanic Gardens at Kew, was commissioned to identify Darwin's *Beagle* plant specimens. They began to correspond, and for the next forty years Hooker was both a sounding board for Darwin's ideas and a vital link to the circles of explorers, diplomats, colonial settlers, and amateur and professional scientists who supplied the now sedentary Darwin with information on the plants, animals and peoples of the world. Hooker was one of the few with whom Darwin shared a theory he had been developing since the late 1830s, one that would account for the diversity of all organic life, and that implied descent of all living things from a single common ancestor.

During the *Beagle* voyage, Darwin had become increasingly aware of the difficulty of distinguishing what were merely varieties from what were different species, of the puzzling similarity between many long extinct fossils and living creatures, and of the very precise adaptation of many organisms to their environment. The idea that species could change over time was controversial but not new; the French biologist Jean-Baptiste Lamarck had proposed that useful characteristics acquired by one generation could be passed on to the next. Writers such as William Paley had cited adaptation as evidence for divine design in nature.

Travelling around the world, Darwin was confronted with the enormous natural variability of even clearly related organisms; back in England he sought out dog and pigeon breeders, observing the dramatic alteration in the characteristics of domestic animals that could be produced in only a few generations by manipulating naturally occurring variations. Thomas Malthus's *Essay on the Principle of Population*, which demonstrated the tendency of populations to grow until limited by competition for resources, gave Darwin the final piece he needed to sketch out a new theory. What he proposed was that any naturally occurring characteristic in an individual – whether plant, animal, bird or human – that helped them survive long enough to breed would be passed on disproportionately to subsequent generations. This principle of 'natural selection' held good for the entire range of physical characteristics – whether coloration for camouflage, the means to fight or take flight, or the ability to access food supplies that others could not exploit. By this mechanism, Darwin realized, whole populations could become very precisely adapted to local conditions, as had a number of bird species he collected from different islands in the Galapagos, most famously the finches – although it was only later work by John Gould that made this clear. Given time enough, offspring of such organisms could diverge, occupying different environmental niches and evolving into new species. Although the word 'ecology' did not exist in English until 1876, Darwin was already in many ways an ecologist.

As with geology, so with biology: once he had the outline of his global theory, Darwin's entire research programme was informed by it. Even his eight-year taxonomic study of living and fossil barnacles, finally published in 1851, is, when seen in the light of his theory, a study of adaptation. As he worked on barnacles Darwin was writing a never-to-be-published 'big book' he called *Natural Selection*, and amassing data on every aspect of natural history, principally through his growing global network of correspondents. One such was Alfred Russel Wallace, a naturalist and commercial collector who by the end of 1856 was providing Darwin with bird specimens from Malaysia. The two men were aware that they had similar ideas. Darwin had told Wallace that he planned to publish on the species question eventually: in addition to several chapters of his big book, he had two earlier manuscripts, one written in 1842 and another in 1844, setting out almost the entire theory. In 1858, Wallace precipitated the publication of Darwin's theories when he sent him a manuscript postulating an identical mechanism for species change.

Distressed to find he might be beaten into print, but distracted by the serious illness of two of his children, Darwin forwarded Wallace's manuscript, as requested, to Charles Lyell, and it was Lyell and Hooker who arranged for Wallace's paper to be read together with one hastily written by Darwin, at a meeting of the Linnean Society. Darwin, distraught at the death of his infant son, was not there. The papers went almost unremarked, but Darwin followed up swiftly, expanding his earlier manuscript with material from the big book, and wrote *On the Origin of Species by Means of Natural Selection* in less than a year. It was published on 24 November 1859 and sold out instantly. Darwin, unlike Wallace, had the social status, scientific credentials and breadth of knowledge to ensure the book was taken seriously. It was also a huge commercial success, being published in six editions during Darwin's lifetime and translated into many languages.

Inventive experiments and enduring fame

Darwin's career was by no means over with the publication of *Origin*. Although suffering from repeated bouts of chronic ill health, he was just fifty, and most of his publications lay ahead. He still lacked a theory of how inheritance worked. Gregor Mendel developed his theory of what we know now as genetics in the 1860s, but his papers remained obscure until after Darwin's death and Darwin never knew of them. A decade after *Origin*, he published a dense two-volume work, *The Variation of Animals and Plants under Domestication*, which concluded with the description of a hypothetical mechanism of inheritance that Darwin called 'pangenesis': he proposed that particles or 'gemmules' circulating in bodily fluids and transmissible from parent to child acted as catalysts for the development of particular organs. Few were convinced, however, and experiments in blood transfusion carried out by his cousin Francis Galton produced no supporting evidence.

His next two books, *The Descent of Man, and Selection in Relation to Sex* and *The Expression of the Emotions in Man and Animals*, were designed to demonstrate how such apparently unique human traits as aesthetic sense, conscience and even religious sentiment were part of an evolutionary continuum with the rest of the animal kingdom. The mechanism of natural selection was supplemented with a theory of sexual selection – organisms had not only to live long enough to breed, but also to possess the characteristics that attracted a mate, and those characteristics too would be disproportionately passed down to future generations.

Darwin's chief preoccupation for much of this time, however, was not with humans, or even with animals, but with plants. Just as he saw humans as merely a strongly marked variety of primate, he saw no hard boundaries between animals and plants. He conducted inventive experiments in his garden and hothouse at Down, investigating adaptations that ensured cross-breeding and therefore greater variation. He studied plants that exhibited animal-like behaviour – plants that moved or were sensitive to external stimuli: the climbers, twiners and insectivorous plants.

Although sometimes criticized later in life for the apparently amateur nature of his working methods, Darwin was a skilled experimenter, and in no way isolated from advances in science. Through his botanist son Francis, who conducted experiments on his behalf, he was in touch with the emergent university laboratories in Germany, and he studied the importance of habitat by arranging for seed sent from one part of the world to be grown in others. He pioneered the use of scientific questionnaires, was a signatory to a number of parliamentary petitions that affected science in public life, and was a quiet supporter of properly controlled vivisection.

Typically, Darwin's last book, published in 1881, *The Formation of Vegetable Mould through the Action of Worms*, was a return to a line of enquiry he had first started decades before, and was the result of experiments carried out with his children in the grounds of his home. It was also the first work to demonstrate the importance of these apparently insignificant creatures to the wider environment and the whole cycle of nature. Darwin died the following year, having seen it outsell all of his previous books. In recognition of his revolutionary scientific ideas and his public fame, he was honoured with a major ceremonial funeral and burial in Westminster Abbey in London.

Gregor Mendel

The founder of genetics and the law
of biological inheritance

(1822–1884)

The distinguishing traits of two plants can, after all,
be caused only by differences in the composition and
grouping of the elements existing in dynamic
interaction in their primordial cells.

GREGOR MENDEL, 'VERSUCHE ÜBER PFLANZENHYBRIDEN'
(EXPERIMENTS ON PLANT HYBRIDS), 1866

In a walled garden of the Augustinian monastery of St Thomas, close to the Moravian city of Brünn (now Brno, Czech Republic), the abbot arranged in 1856 for the construction of a greenhouse for experimental plant breeding. Gregor Mendel, one of the younger members of his teaching community, who had returned from studies at the University of Vienna, was to make famous use of this facility and the experimental garden beside it. During the period between 1855 and 1863, he discovered the scientific basis of biological inheritance, as expressed in the formation and development of hybrids between varieties of the garden pea *Pisum sativum*. Nine years of experimentation, involving more than 10,000 plants, carefully examined, saw him defining a theory based on elements, later to be known as genes, within the pea's reproductive cells, controlling the *Anlagen* ('foundations') for the particular traits he was investigating.

Mendel was born and spent his childhood in the village of Heizendorf (now Hyn ˇcice) in the north-eastern corner of Moravia, the only son of a gardener's daughter and an ex-soldier of the Austrian army. They were peasants struggling to make a living out of a small-holding but determined for their studious son to rise beyond their own status through education. Though it was difficult to find the school fees, they managed to support him at the Gymnasium at Troppau (now Opava) until he was sixteen. After that he financed himself by giving private tuition, which became critically difficult when he progressed, aged eighteen, to the Philosophy Institute at Olmütz (now Olomouc), a preparatory college of the university. There he struggled against ill health, forced to return home several times to recover. In pity, his younger sister gave him part of her dowry. Despite inter-ruptions to his studies, Mendel's academic quality was recognized at the Institute, setting him on course, at the age of twenty-one, to entering a religious order in which his scholarly ambitions would be strongly encouraged, and education provided at the highest level. His fascination with science would take him in various directions, including meteorology and bee culture, as well as heredity in plants, all of which he combined with full-time teaching duties in physics and natural history at Brünn's Oberrealschule.

His choice of the garden pea as experimental material rested on its being normally self-fertilized, allowing the isolation of pure breeding varieties, immune to accidental cross-pollination because of the keel-shaped form of the pea flower, which tightly enclosed the reproductive structures within. In the report of his discovery published in 1866, Mendel explained that before starting the experi-ments he had made sure that all the traits he wished to study were consistently expressed, by growing thirty-four varieties side by side for two years. To make crosses between different varieties he used 'artificial fertilization': the removal of male organs from flowers (to prevent self-pollination) followed by the introduction of pollen from another variety. His aim had been 'to obtain new variants' through further understanding of 'the development of hybrids in their progeny'.

Horticulturalists who had crossed pea varieties before had remarked on the unpredictable burst of variation in the second-generation (known as F2) hybrids, to which Mendel responded by analysing the variation statistically and explaining it mathematically. It was through his unique *statistical* approach to variation that he revealed his originality as a scientist.

Experiments in a monastery garden

Mendel's two years of intensive study at the University of Vienna prepared him to take account of the modern opinions of 'famous physiologists' that 'propagation in phanerogams (flowering plants) is initiated by the union of one germinal and one pollen cell into a single cell'. His rejection of the traditional idea of the pollen cell alone being the origin of the embryo was critical to his analysis of variation. It was the very basis on which he advanced his study of the differences between hybrid generations. As an accomplished field botanist he was used to distinguishing one species from another on the basis of contrasting traits. In evaluating differences between generations of hybrids, he followed a similar procedure, separating individuals in terms of defined pairs of 'contrasting characteristics' (traits) of colour, size and shape to establish whether they were in any way associated or varied independently.

Observing how every one of the seven pairs of contrasting pea characteristics that he investigated separated ('segregated') and recombined through successive generations, Mendel showed that, for all traits the first-generation (or F1) hybrids were uniform in resembling one member of each contrasting pair of traits (for example, all F1 plants from a cross between green- and yellow-seeded varieties bore only yellow seeds). He defined this as the dominant character, observing that dominance was the same irrespective of the direction of the cross (yellow female × green male, green male × yellow female). His originality was demonstrated by what followed next.

He examined the F1 progeny (F2) and found that, in all cases, one of the traits of each contrasting pair proved much more common than the other. Counting them in large samples, he discerned that for every pair, the dominant trait exceeded its alternative recessive partner in a ratio approaching closely to 3:1 (3 yellow to 1 green in the example quoted). Mendel explained this pattern by a well-known mathematical principle. Representing the dominant trait as A and the recessive contrasting trait as a, the first-generation hybrid would be Aa, which when interbred ($Aa \times Aa$) would yield the combination series $AA + 2Aa + aa$ in F2. Since A was dominant to a, the 3:1 ratio in the second-generation hybrid was explained.

Having attributed the changes to a mathematically predictable process, he carried out a series of experiments to establish that every possible question about his theory was answered. In probing the origin and development of hybrids arising from his crosses, through generations where there were two or three pairs of traits, he discovered patterns both of heredity (that is, constancy between generations) and variation in the interrelationships of separate traits. After also investigating backcrosses, he was able to claim his theory to be well proven. All seven traits were assorting independently as they passed through the generations. With a physicist's understanding of probability and combination theory, he had traced through nine generations of crosses the passage of traits both individually and in association with one another. He had started his experiments 'to deduce a law' (his term). By the time of publication, he was convinced of its reality.

A theory too innovative for acceptance

Mendel's theory, published in 1866, was widely circulated internationally but met with no understanding. He had turned an aspect of natural history into a science with the unique property of explaining heredity and variation as two aspects of a single developmental process, resulting from the formation of the reproductive cells (female and male) and their coming together at fertilization. But no fellow scientist admitted to sharing his enthusiasm. As Mendel considered his traits individually and in combination, he could make predictions about 'the degree of kinship between hybrid forms and their parental species'. This was because his experiments had demonstrated that the inherited elements passed unchanged between generations. He wrote to Karl Wilhelm von Nägeli, the famous Swiss botanist, that this had been 'precisely determined' from his experiments. In a further letter he reemphasized this point, stating that for segregating traits 'there is no indication that one of them has either inherited or taken over anything from the other'. He also mentioned that he had used this knowledge to practical advantage in isolating a very fertile variety of peas with large, tasty seeds. But Nägeli did not appreciate the value in Mendel's theory, nor until 1900 did any other scientist.

Two years after Mendel had revealed his theory to a world unready for his findings, the monastery suffered the loss of its revered abbot. Mendel's election to the abbacy gave him many extra responsibilities to divert him from his garden. Further hybrid studies came to an end in 1871. But he treasured his discovery and continued to apply its principles to practical benefit. This is known from the report of a visit from a young horticulturalist from France one summer's day in 1878. As the two of them walked round the grounds together, Prelate Mendel showed him many labelled varieties of fruit trees, hot-house plants and well-grown vegetables, including 'several beds of green peas in full bearing, which he said he had reshaped in height as well as in type of fruit to serve his establishment to better advantage'. When asked how he had achieved this, Mendel replied: 'It is just a little trick.'

Genetic inheritance

Investigations since the 'rediscovery' of Mendel's work in 1900 have revealed the essence of his 'little trick' and its predictive qualities. The history of genetics has been characterized by the successive ways in which his concept of the inherited element (gene) has been redefined, both in material terms and theoretically. The fundamental significance of Mendel's discovery is now universally acknowledged. The achievement of this most modest of great scientists, in the service of his monastery and in answer to his own curiosity, created what today we call genetics.

Jan Purkinje

Investigator of vision and pioneer of neuroscience

(1787–1869)

*There are sensations that do not correspond to anything
outside the body. In so far as they imitate the qualities and
forms of external things, they thereby often give rise to
illusions, phantoms, or appearances with no corresponding
reality. These can be referred to as subjective sensory
phenomena. Yet it remains the undeniable task of the
natural scientist to establish their objective basis
while for the general use it is enough to know that
they involve only the sensory organs and that
we need not search for external objects.*

JAN PURKINJE, *CONTRIBUTIONS TO THE KNOWLEDGE
OF VISION IN ITS SUBJECTIVE ASPECTS*, 1823

Neuroscience, as a discipline, did not exist in the 18th and 19th
centuries. It emerged as a consequence of the endeavours of many
who conspired to illuminate the structure of the nervous system,
the manner of communication within it, and its links to perceptual
experience. Jan Purkinje, one of the best-known scientists of his time,
was present at the dawn of neuroscience and he added greatly to the
structures, processes and phenomena that still reside at its heart.
Indeed, the heart itself was subjected to his scrutiny, as is attested by
the Purkinje fibres surrounding it. Neuroscience emerged from the
biological sciences because conceptual building blocks were isolated,
and the ways in which they can be arranged were explored. Purkinje
was not trained in histology – the study of the microscopic anatomy
of cells – but he became father of the discipline.

Jan Evangelista Purkinje was born in Libochovice, Bohemia, and his native language was Czech. Nine different spellings of his name have been found, but he used the German version Purkinje in most of his publications, reverting to the Czech, Purkyňe, after his return to Prague in 1850. Purkinje's work has fascinated students from many disciplines, but he is best known for his early descriptions of cells, some of which bear his name: the Purkinje cells in the cerebellum, for example. He is recognized by physiologists, who admire his attempts to relate structure to function; by pharmacologists, who marvel at his heroic experiments with self-administered drugs; and by forensic scientists, who acknowledge his role in the use of fingerprints for identification. Yet all of these achievements followed his initial enquiries into vision. Purkinje began experiments on his own sight when he was a student in Prague, and he assembled and extended the observations in his dissertation for the degree of doctor of medicine; he sought to describe a range of subjective visual phenomena and to account for them in objective terms.

Discovery based on visual phenomena

Vision provides a bond that unites physiology and psychology, and it is this bond that was strengthened by Purkinje's enquiries. He was an observer *par excellence*, and several visual phenomena are named after him: the Purkinje shift, Purkinje images and the Purkinje tree. These were described in his books of 1823 and 1825 on subjective visual phenomena, which can be thought of as defining a new area of enquiry. He did not see further than others by standing on the shoulders of giants; rather he looked inwards and discovered a visual world that is still being explored to this day.

In 1823, Purkinje was appointed professor of physiology at the University of Breslau (present-day Vroclaw in Poland, but then located in the German state of Prussia). In his inaugural lecture, he described the images that are reflected from the surfaces of the cornea and lens; these are now called Purkinje images, and have been used to determine the mechanism of accommodation – that is, of focusing – as well as providing a technique for monitoring eye movements. He also illustrated the nine main patterns of fingerprints, and the principles upon which the ophthalmoscope is based. His experiments on vertigo laid the foundations for subsequent vestibular research: noting the way the eyes moved after body rotation and linking it to head position enabled later researchers (such as Ernst Mach, Josef Breuer and Alexander Crum Brown) to formulate the hydrodynamic theory of semicircular canal function within the inner ear, which senses movement and helps our balance and spatial awareness.

In visual science, Purkinje is perhaps best known for identifying the brightness changes of colours at dawn or dusk: blue objects that appear brighter than red ones before sunrise reverse thereafter. He described it thus: 'The degree of objective illumination has a great influence on the intensity of colour quality. In order to prove this most vividly, take some colours before daybreak, when it begins slowly to get lighter. Initially one sees only black and grey. Then the brightest colours, red and green, appear darkest. Yellow cannot be distinguished from a rosy red. Blue looks to me the most noticeable. Nuances of red, which otherwise burn brightest in daylight, namely carmine, cinnabar and orange show themselves as darkest, in con-trast to their average brightness. Green appears more bluish, and its yellow tint develops with the increasing daylight.' This phenomenon is now called the Purkinje shift, and it can be related to the different spectral sensitivities of rod and cone receptors in the retina: rods are more sensitive to shorter wavelengths of light than cones are. He also constructed a perimeter to explore the colour zones of the retina. The Purkinje tree refers to the visibility of blood vessels in the retina when the eye is illuminated from the side.

In 1832 Purkinje obtained an achromatic microscope made by Georg Simon Plössl, one of the best microscopes in the world at that time. He directed it at the cerebellum, thereby identifying the cells that bear his name. He was also one of the first to use a microtome, a mechanical device for slicing thin tissue sections for microscopic examination. Purkinje's laboratory at Breslau has been described as the cradle of histology, not only on the basis of his microscopic observations, but also because of the ways in which he taught and encouraged his students to learn through discovery. Such an approach was not without its opponents. When Purkinje introduced demonstrations and laboratory work to his teaching classes, the university faculty sought to have him demoted. He had influential supporters, however, including the poet Johann Wolfgang von Goethe, and the Prussian ministry of education commended his methods. He went on to overcome the initial hostility, won the respect and friendship of his colleagues, and became one of the best-known teachers at the university. In November 1839, Purkinje established the world's first independent department of physiology, and three years later he opened the Institute of Physiology, the world's first official physiological laboratory.

In 1850 he returned to Prague to establish an institute of physiology. His research was published mainly in the Czech language. He founded and edited journals, and he promoted the value of education. He also entered into Czech cultural life and played an important role in the national revival. In his last decade, Purkinje had many international honours bestowed upon him, and in the year before his death he was given the Order of Leopold, one of the highest honours of the Austrian Empire. He died in Prague.

Santiago Ramón y Cajal
The fine structure of the brain
(1852–1934)

On a perfectly translucent yellow field appear thin,
smooth, black filaments, neatly arranged ... arising from
... black bodies! One might say they are like a Chinese ink
drawing.... The eye is disconcerted, so accustomed is it
to the inextricable network stained [in the usual way]
which always forces the mind to perform feats of critical
interpretation. Here everything is simple, clear,
without confusion.

SANTIAGO RAMÓN Y CAJAL, *HISTOLOGIE DU SYSTEME NERVEUX*
DE L'HOMME ET DES VERTEBRES, 1909

Thus did the older Santiago Ramón y Cajal describe his moment
of epiphany as a young anatomist in 1887 when he first saw nerve
fibres stained in the new way described by the Italian scientist and
physician Camillo Golgi. Cajal had been born in a small village in the
Spanish countryside on 1 May 1852, the son of a barber-surgeon who,
through great effort, later obtained a medical degree and demanded
that his son also attend medical school. But like many gifted young-
sters, Cajal was a rebellious teenager, reacted poorly against the rigid
discipline of his schooling, showed artistic talent early and intended
to become an artist. This appalled his father, who eventually won
out, and Cajal graduated from the medical school in Zaragoza in 1873.
He promptly joined the army medical service, contracted malaria in
Cuba, and was shipped home after less than a year. Because anatomy
had been the only subject for which he showed much interest or

14 Linus Pauling stands next to a model of collagen at the California Institute of Technology, Pasadena, in 1958. Determining the precise molecular structures of various substances was a principal goal in much of Pauling's scientific work, and he often constructed models, such as the one behind him in this picture, to illustrate the bond lengths and bond angles of these substances.

15 The technology behind high-energy particle physics. The photograph shows the magnets and distribution feedbox in the final focus of one of the CERN laboratory Large Hadron Collider's six detectors, located in a 17-mile (27-km)-long tunnel 500 feet (150 metres) beneath the Franco-Swiss border.

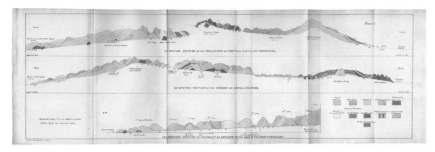

16 Ten years after returning from the *Beagle* circumnavigation, Charles Darwin published his *Geological Observations of South America* (1846). This included ambitious geological cross-sections through the Andes to support his theory of large-scale subsidence and uplift. The printed versions were based on a series of hand-drawn and coloured sections that the young Darwin had literally pieced together during the voyage on bits of paper glued together in long strips.

17 James Watson (left) and Francis Crick photographed with their model of DNA at the Cavendish Laboratory, Cambridge, a few weeks after they published a paper announcing their discovery in *Nature* magazine on 25 April 1953.

18 A Purkinje cell (in green) exposed by a modern staining procedure. One of the ironies of science is that Jan Purkinje was able to see only the large cell body and not the dense arborizations of a single Purkinje cell because of the staining techniques and microscopes available in the 19th century.

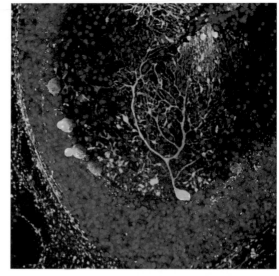

Heredity in *Primula Sinensis*.

1. Primrose Queen. 2. Crimson King. 3. F₁ formed by crossing these two types. 4—21. Various F₂ types obtained by self-fertilising F₁. 4, 10, 16. Whites. 5, 11, 17. Various tinged whites. 6, 12, 18. Light magentas. 7, 13, 19. Reds. 8, 14, 20. Magentas. 9, 15, 21. Deeper magentas. 7, 13, 15, 19, 20 have the dark blotches which cannot appear unless the stigma is red. 16—21 are all large eyed, viz. homostyle forms, like 1.

19 Plate from *[Gregor] Mendel's Principles of Heredity* by William Bateson, published in 1909, showing how unexpectedly wide variation in an inherited trait can arise when two or more separately inherited elements interact, as in the case of *Primula sinensis*, or Chinese primrose, where one parent has white flowers and the other, dark red. Variation in the size and shape of the flower and the characteristics of the 'eye' are explained by the actions of other segregating elements.

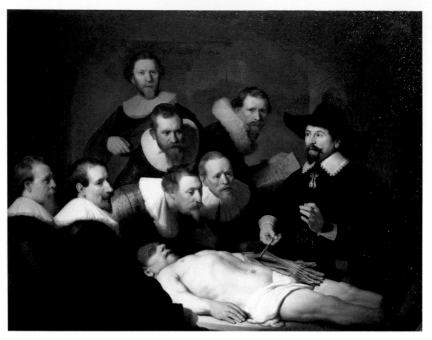

20 Rembrandt van Rijn's *The Anatomy Lesson of Dr Nicholaes Tulp* (1632) featured the newly appointed public anatomist of the Amsterdam Guild of Surgeons, wealthy members of the guild who had paid to be included in the portrait, and the corpse of Aris 't Kint, executed for the theft of a coat before the final punishment of dissection.

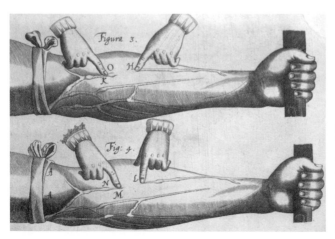

21 Plate from William Harvey's *De motu cordis* (1628), about the circulation of blood, showing the operation of valves in the forearm.

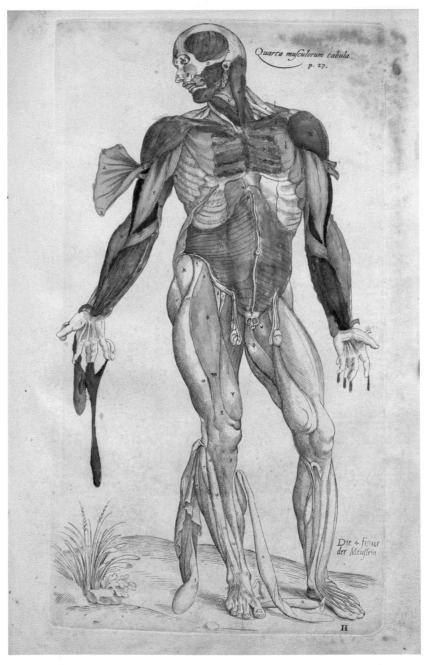

22 A page from Andreas Vesalius's *De humani corporis fabrica* (On the structure of the human body), the foundational text of modern human anatomy, first published in 1543.

23 Leakey expeditions were family affairs. Here, Louis, Mary and their eleven-year-old son Philip excavate a nearly two-million-year-old early human occupation site in Tanzania's Olduvai Gorge in 1960 – with the assistance of their dalmatians and fox terrier.

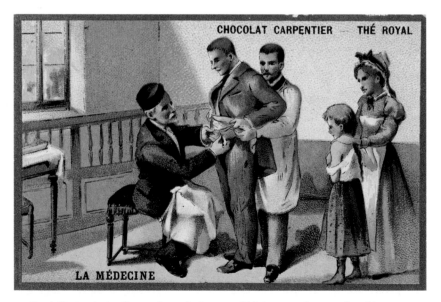

24 Louis Pasteur's standing and popularity caused his image to be reproduced in many forms of 19th-century popular media, including on chocolate boxes. This illustration shows a dog-bite victim being inoculated against rabies by Pasteur; in fact, he never treated a patient himself because he was not medically qualified.

25 Alan Turing, mathematician, code-breaker and father of modern computing; one of a set of eight different undated passport-sized photographs of from the archives of King's College, Cambridge.

26 J. Robert Oppenheimer (left), former technical director of the Manhattan Project, and John von Neumann stand in front of Princeton's IAS machine in 1952. Built from late 1945 until 1951 under von Neumann's direction, the machine was the first electronic computer constructed by the Institute for Advanced Study in Princeton. It could perform 2,000 multiplications in one second and add or subtract 100,000 times in the same period.

27 Sigmund Freud in 1922. Even if many of his ideas were wrong, Freud still caused a revolution in the way we think about our selves – our bodies and our minds – that continues to provoke and intrigue neuroscientists.

aptitude, he then embarked on a career that depended a great deal on his artistic talents. It took him to academic posts of increasing prestige in Valencia, Barcelona and Madrid, winning him international recognition for his discoveries in the cellular architecture of the nervous system, and a share of the Nobel Prize in Physiology or Medicine with Golgi in 1906.

Seeing inside the brain

Until the mid-19th century, the use of the microscope to study cells and tissues was limited by primitive techniques. This became a special problem as one studied the nervous system. The 'cell theory' – that all living bodies were composed of cells – was enunciated in 1839 and promptly accepted as applying to all organs, with the lone exception of the nervous system. Yes, the nervous system operated by electricity, but how was this electricity generated, how was it transmitted, and how could this be related to its fine structure? Even the brain and spinal cord seemed to be made up mostly of fibres. Although cells were seen, they could not be made out very well amid the mass of fibres, and it was uncertain whether they served only the secondary purpose of nourishing the fibres or whether they had some greater importance. Golgi provided a solution by discovering a method of staining nervous tissue with silver chromate so that a limited number of cells could be stained in their entirety, enabling paths of nerve cells in the brain to be seen for the first time.

But the Golgi stain could not be depended upon, was difficult to reproduce, and was not much used in the period 1880–85 when Golgi reported on it. He became so discouraged that he turned away from neurohistology to research on malaria, a study that made him famous. Cajal took over and made improvements in the staining process and proceeded over the years to make a host of original discoveries. He observed a multitude of different types of nerve cell – now known as neurons – and how they related to each other in different parts of the central nervous system. The depth and breadth of his observations were such that current neuroanatomists still invoke him when they write of their own findings today.

Certain features of Cajal's character were especially well suited to the demands required to accomplish this. He had the imagination to pick important problems to study and spot exactly where in the central nervous system to study them. Because he could rarely, if ever, see an entire neuron in one field of view, he had patiently and tenaciously to trace it through many levels in different views. He had then to synthesize details of these views and render what he saw accurately and meticulously. He had no peer in grasping the interrelationships between neurons rarely seen together in a single view through the microscope. His forte was not in tracing long tracts through the nervous system but rather in appreciating the connections between cells in particular regions, such as the cerebellum, cerebral cortex and retina. He was not above drawing a space between the fibres of one neuron and that of a contiguous one in order to emphasize the discontinuity he felt was there, even though such a space was invisible to him.

A key founder of modern neuroscience

His studies were important in establishing the main principles we accept today as basic to understanding the nervous system. First, the neuron is the basic unit of the system. The shorter dendrites and the longer axons are outgrowths from the cell body, which provides nourishment to these elongated processes. Secondly, nerve impulses travel almost always in only one direction in a neuron, from the dendrites toward the cell body, then through the cell body to the axon and its branches, which in turn make connections to the dendrites or cell body of another neuron. Thirdly, each neuron maintains its individual integrity. There is a physical discontinuity between neurons: the connections do not involve any physical merging with the processes of another neuron. The nerve impulse travels in only one direction across this discontinuity (later to be termed a 'synapse'); nerve pathways are thus chains of neurons and their synapses.

Cajal was an uncompromising advocate of what became known as the neuron theory, a concept that had autonomous neurons following circumscribed conduction pathways through the nervous system. Only this, he felt, could account for the rapidly emerging evidence for the localization of specific functions in specific places in the brain and spinal cord. Especially useful were his meticulous studies of cells in the cerebral cortex. They merged with the parallel efforts of several others to provide a basis for mapping vision, touch and motor representation in the discrete locations in the cortex. The neuron theory proceeded to have such great explanatory and predictive power in the physiology and pathology of the 20th century that it became an accepted foundation of modern neurobiology, with Cajal himself regarded as a major founder of modern neuroscience. During his lifetime, however, the neuron theory had its critics. Cajal was still defending it before his death in 1934. Not until the 1950s was the theory finally proven by the visualization of the synapse through the electron microscope.

Francis Crick and James Watson

Decoding the structure of DNA
and the secret of life

(1916–2004 and born 1928)

I'm Watson, I'm Crick,
Let us show you our trick,
We've found where the seed of life sprang from,
We believe we're a stew
Of molecular goo
With a period of thirty-four Ångstrom.

DOGGEREL ACTED BY ROY MARKHAM AND E. S. ANDERSON
AT THE COLD SPRING HARBOR SYMPOSIUM, 1953

The names Crick and Watson will forever be associated with the three letters 'DNA', standing for the chemical deoxyribonucleic acid – in other words, the genetic material. They did not discover this substance. That achievement belongs to the previous century. What they discovered on the last Saturday in February 1953 was its molecular structure. It is composed of two helical chains, forming a cylinder, the flat bases piled in pairs inside like piles of pennies.

This was one of a number of successes achieved in the 1950s and 1960s by scientists attempting to solve the structures of complex molecules. The American chemist Linus Pauling had provided the initial stimulus to such efforts with his suggested structures for fibrous proteins in 1951. But of all the successes of those years, it is the solution of the structure of DNA that stands out, for it had the greatest impact on the whole of biology. Why was this so? Because it is the very material of which genes are made, and its structure is at once so suggestive of the many ways in which it performs its functions: the duplication, transmission, mutation and expression of genetic information. A revolution in biology was the result. It has spread from biophysics to biochemistry, genetics, virology, the medical sciences and beyond. Called 'molecular biology', this research tradition has acquired an army of techniques for understanding and manipulating DNA. As a result, the determination of the sequence of the four bases of DNA has been achieved. Within these sequences lie the messages of the genes.

Francis Harry Compton Crick was the firstborn of Harry and Annie Crick. They lived in Northampton, where his father ran the family footwear company. There were no academic influences within their circle, but Francis displayed an interest and aptitude in science from an early age, and at seventeen won his school's chemistry prize. He went on to study physics at University College London, gaining his degree in 1937. His record there was only average, but the need for physicists in World War II brought him to a research position with the Royal Navy. He distinguished himself in the Mine Design Department, devising novel firing mechanisms to apply to mines. He displayed an originality, a shrewd judgment and, behind his boyish exterior, a forceful personality and very sharp mind. The reputation thus gained brought him strong support for his request in 1947 to move into biophysics. His new career began at the Strangeways Laboratory in Cambridge, but in 1949 he joined the Medical Research Council Unit, housed in the Cavendish Laboratory under the direction of Max Perutz, and began to study protein structure.

James Dewey Watson was brought up on Chicago's South Side. His father was a bill collector and his mother, a company secretary and accountant. They were not wealthy, but Jim was a bright lad, and at fifteen years of age he gained entry to the University of Chicago's four-year college programme. By the age of nineteen, he entered graduate school at the University of Indiana to research the action of X-rays on bacterial viruses, completing his doctorate in 1950. He spent the next couple of years as a postdoctoral researcher in Europe. Early in October 1951, he moved from Copenhagen to Perutz's unit in Cambridge to get to grips with X-ray crystallography, in the hope of finding the structure of DNA. It was at this point that he was introduced to Crick. Watson recalled how on that day, 'I knew I would not leave Cambridge for a long time. Departing would be idiocy, for I had immediately discovered the fun of talking to Francis Crick.' Years later he described that first encounter: 'I witnessed Crick's irrepressible brain in action ... like watching Fourth of July fireworks. Never before had I witnessed such disciplined intelligence.'

A perfect match of opposites

And yet what a contrast in personalities they represented. Crick, aged thirty-five, was voluble, loquacious, ebullient and supremely confident; he was direct as he greeted you, and his searching blue eyes looked straight at you. Watson, his speech irregular, often interrupted with a kind of snigger, would glance awkwardly this way or that as he addressed you. But aged only twenty-three and already with a PhD, he was clearly well ahead of his contemporaries. Twelve years his senior, Crick still had no doctorate, but that was due to his war work.

Despite these differences, they proved ideal collaborators because Crick, although working in the laboratory on an experimental project, loved to talk, discuss and debate. Subject matters abounded, for his appetite for scientific literature was voracious. And when Watson was around, he could talk to Crick about his mission to discover the structure of DNA. It was Watson who brought the importance of DNA to the attention of the MRC Unit and convinced Crick that nucleic acids were probably the chief, and possibly the only, hereditary material. So when Crick was busy in the lab with proteins, Watson would be searching the literature on nucleic acids in the library. Then they would discuss DNA, and Watson would inform Crick about the latest news from the 'phage' group in America. Constituted of geneticists and former physicists, this group was researching the genetics of bacterial viruses (or 'phages'). Their minuteness and the rapidity of their reproduction gave geneticists the hope to advance their analysis of the genetic map close to the molecular level.

The discovery of DNA's structure

Watson had not long been in Cambridge before he and Crick built a theoretical structure for the DNA molecule. It was wildly wrong, a complete disaster, causing the Cavendish Professor, Lawrence Bragg, to put a moratorium on DNA research, thus avoiding duplication of work already ongoing at the other MRC Unit in London. However, in February 1953 Bragg permitted Crick and Watson to make a further attempt, because Pauling was about to publish a structure for DNA. To Bragg that meant 'open season' had arrived. By the end of the month, they had made a breakthrough. Watson later offered a vivid account of the events leading to the pair's second structure for DNA. Constructing the two sugar-phosphate backbones of this long helical molecule, giving the sugar residues the right angular displacement and running the chains in opposite directions were Crick's contribution. Pairing the bases – adenine (A) with thymine (T) and guanine (G) with cytosine (C) – was Watson's. The latter was suggestive of the manner in which DNA might duplicate, for if the two 'mother' chains could be separated, free bases could attach to the bases of the single strands produced, yielding two double DNA daughter molecules where there had been but one.

Crick and Watson immediately realized their structure was hot news, although it would take four to five years before the scientific community at large began to have confidence in their model. But the two men were later to marvel at the extent of the revolution it had produced.

Divergent career paths

Before the discovery of DNA structure, Crick had been expecting to leave the MRC Unit; Bragg, he knew, found his presence irksome. Accordingly, when invited in late 1952 to join the Protein Structure Project in Brooklyn, New York, for the following academic year, he had accepted. He completed his PhD thesis in the summer of 1953, and he and his family left Cambridge for Brooklyn, where he would be working on the structure of the protein ribo-nuclease. Returning to Cambridge in September of the next year, Crick did not immediately turn back to the structure of DNA. It had been the task of his friend Maurice Wilkins to confirm or modify Crick and Watson's model. (For that work, Wilkins was jointly awarded the 1962 Nobel Prize in Physiology or Medicine along with Watson and Crick.)

In the winter of 1955/56 Crick and Watson collaborated once more, choosing this time virus structures as the subject of their research. But then Watson left Britain and returned to the United States to become assistant professor at Harvard University's biology department. For his part, Crick was set on a career in research, and to this end he avoided all possible administrative and teaching responsibilities. Behind the scenes, however, he exerted considerable influence on the future development of the MRC Unit as it became the Laboratory of Molecular Biology in 1962. His influence was strong too in the planning of the Salk Institute for Biological Studies in La Jolla, California, of which he became a non-resident fellow in 1961, with research programmes in experimental biology and the neurosciences. His relationship with the University of Cambridge and its colleges was not close. Although he became a fellow of Churchill College in 1960, he resigned a year later over the decision to build a chapel. Subsequently he accepted honorary fellowships at Churchill College and at Gonville and Caius College.

Meanwhile, Watson was identifying talent and coaxing outstanding young researchers such as Walter Gilbert and Mario Cappechi to join his research team at Harvard. His 1965 textbook *The Molecular Biology of the Gene*, based on his lectures to introductory biology students, was the first of its kind; it is still in print and now in its sixth edition. Then he began writing his recollections of how he and Crick discovered the structure of DNA. Despite Crick's and Wilkins' efforts to prevent its publication, the book appeared in 1968 under the title *The Double Helix*. Called 'fresh, arrogant, catty, bratty, and funny', it became a bestseller, making Watson a celebrity. Crick, who after winning the Nobel Prize had sought to avoid such a fate, now found his task more difficult. *The Double Helix* drew attention to DNA and to the extent of the two discoverers' dependence on the fine work of the British X-ray crystallographer Rosalind Franklin, working at King's College London. Her early death four years before the award of the Nobel Prize in 1962 meant that her contribution could not be acknowledged.

Towards the genetic code?

From the outset, Crick and Watson were convinced that the structure they had identified, if correct, would show the way to an understanding of how genes encode hereditary information and how their products – enzymes or antigens – all of them proteins, determine the characteristics of an organism. Before the pair's discovery it was an open question just how the detailed instructions for building an organism could be encoded in its DNA. But their model was suggestive, for the four bases – A, T, G and C – could be arranged on one chain of the double helix in any imaginable sequence, and the base-pairing rules discovered by Watson would then dictate the base sequence on the complementary chain. Such a four-letter code could easily specify one of the twenty 'primary' amino acids that living cells use to encode proteins. But what are the salient features of that code? And how is the genetic language of the nucleic acids translated into the protein language of the gene product?

It seemed that ribonucleic acid (RNA) played a part, for it appeared to travel from the cell's nucleus to the surrounding cytoplasm where it was associated with protein synthesis. Could discovering RNA's structure offer clues to the genetic code? Crick certainly thought so. Hence his efforts with Alexander Rich at the California Institute of Technology (Caltech) and others to find it. Watson, too, hoped in vain for clues from RNA, but it would not reveal its structure. Others put their hopes in combinatorial mathematics. In 1957 Crick's productive collaboration with Sydney Brenner in Cambridge began, and by December 1961 they were able to publish the general features of the genetic code – but they could not 'break' the code. That was to be the work of various biochemists over the coming decade. Crick followed their progress eagerly, and summarized their results in chequerboard fashion.

New horizons in neuroscience and genetics

In 1976 Crick spent a sabbatical at the Salk Institute. Offered a personal chair there, he accepted in 1977 and resigned from the Laboratory of Molecular Biology. At this point, he switched fields from molecular genetics to neuroscience. His publications in this area began in 1979 and continued until the end of his life, collaborating with Christof Koch at Caltech. He proved a stimulating presence in this young conglomerate of disciplines, and played an important role in attracting neuroscientists to the Salk Institute.

In contrast to Crick, Watson made his mark in science as a writer and an entrepreneur. In 1968 he undertook to rescue the long-established Cold Spring Harbor Laboratory from its probable demise. Under his direction and with his forward vision and fundraising ability, it prospered and expanded, becoming one of the leading biological research institutes in America. He left Harvard University in 1988, and the same year became director of the Human Genome Project, until resigning in 1992. The following year, Bruce Stillman succeeded

Watson as director of the Cold Spring Harbor Laboratory and Watson became president, then chancellor, and finally emeritus in the years 2008–19, before parting ways. Watson received the National Medal of Science from President Clinton in 1997, and an honorary British knighthood in 2002. Crick refused a knighthood but accepted the Order of Merit from Queen Elizabeth II in 1991.

BODY AND MIND

In 1813, the English polymath Thomas Young – a physician who was also a physiologist of the eye, a physicist and mathematician, and a linguist and Egyptologist, among other things – wrote in his *Introduction to Medical Literature*, 'There is no study more difficult than that of physic: it exceeds, as a science, the comprehension of the human mind.'

Although medicine since then has advanced as a science almost beyond recognition, Young's warning is worth remembering in regard to all science of the human body and mind and those who dedicated themselves to it, such as the anatomist Andreas Vesalius, the physician William Harvey, the bacteriologist Louis Pasteur, the psychologist Francis Galton, the psychiatrist Sigmund Freud and the physical anthropologists Louis Leakey and Mary Leakey, discoverers of the fossils of the earliest known hominins – upright, two-legged, early humans – in the rocks of east Africa. It also applies to the computer scientists Alan Turing and John von Neumann, who were interested in human–machine interaction and the possibility of artificial intelligence. The Turing test of machine intelligence is that a computing machine, when kept separate from a person, must appear human when interacting with that person in a natural-language conversation. No computer has passed the test.

Each of these scientists was compelled, to a greater or lesser extent, to grapple with the unresolved philosophical issue raised by Descartes in his famous 'mind–body problem': that psychological phenomena appear to be qualitatively and substantially different from the physical bodies on which they apparently depend. If this dualism is real, as Descartes maintained, can science, in particular neuroscience, ever understand psychological phenomena? If it is not real, then is the self merely an illusion produced by the physical processes in the brain? Vesalius was the first human anatomist to emphasize the necessity of personal, hands-on, dissecting of corpses, rather than placing one's trust in books, notably the classics of the ancient Roman anatomist Galen of Pergamon, whose errors, Vesalius came to realize, derived from Galen's reliance on dissecting animals. While dissecting, Vesalius worked with an artist at his side, and eventually published, during the High Renaissance, his monumental study *De humani corporis fabrica* (On the structure of the human body): seven large folios on the skeleton, musculature, veins and arteries, nerves, reproductive and digestive organs of the abdomen, heart and lungs, and the brain and sense organs, illustrated with seventy-three astonishingly graphic illustrations that changed anatomy for ever. Vesalius did not, however, correctly explain the heart and blood. It was Harvey who, in 1628, overturned the Galenic doctrine that venous blood was generated in the liver, from whence and after absorbing air from the lungs the blood ebbed and flowed through the ventricles of the heart into the arteries. In fact, Harvey proved, the body had a fixed amount of blood that moved around it via the heart in a 'circular' fashion.

Pasteur, in his pioneering work on the stereochemistry of sugars, his studies of fermentation leading to his germ theory of disease, and his development of pasteurization and vaccines against anthrax, chicken cholera and rabies, is perhaps the archetypal 19th-century scientist: rational, driven and public-spirited. 'In the fields of observation, chance favours only prepared minds', he said. Yet, in his famous 1860s dispute over the spontaneous generation of life in ferments, Pasteur opposed spontaneous generation, taking the view that the origin of life was from God and not physical forces.

In psychology and psychiatry, Galton and Freud are both fascinating figures, but controversial too: Galton for his founding of the eugenics movement, and Freud for his founding of psychoanalysis. Galton's ambitious projects to measure intelligence and to show genius to have been largely inherited were stymied by his inability to provide a scientific definition of intelligence and genius – a problem that persists today. Freud's attempts to study dreams, thoughts and feelings and to discover in them mechanisms – for instance, 'repression' – were even more intractable; indeed Freud himself doubted that psychoanalysis should be called a science. But even if many of Freud's ideas were wrong, he still caused a revolution in the way we think about our selves – our bodies and our minds – that continues to provoke and intrigue neuroscientists.

Andreas Vesalius

Renaissance anatomist of the human body

(1514–1564)

If anyone wishes to observe the works of nature,
he should put his trust not in books on anatomy but
in his own eyes and … industriously practice
exercises in dissection.

GALEN OF PERGAMON, *ON THE USEFULNESS OF THE PARTS*,
BOOK 2, CHAPTER 3, 2ND CENTURY AD

In July 1543, Andreas Vesalius published his monumental *De humani corporis fabrica* (On the structure of the human body), known simply as the Fabrica. This comprised seven large folio books (sections) on the skeleton, musculature, veins and arteries, nerves, reproductive and digestive organs of the abdomen, heart and lungs, and the brain and sense organs. The detailed Latin text was accompanied by seventy-three breathtaking, true-to-life illustrations. Anatomy would never be the same again.

Vesalius was born in Brussels, then in the Habsburg Netherlands, the son of well-connected parents: his father was an apothecary at the court of Charles V, ruler of the Holy Roman Empire; his mother was the daughter of a prosperous civil servant. He enjoyed a propitious education at the University of Leuven (or Louvain), benefiting from the progressive humanism of Erasmus of Rotterdam. Erasmus encouraged the study of original texts – which started arriving in Europe from the early Renaissance – in their ancient languages (classical Latin, Greek and Hebrew), rather than using those filtered through generations of Islamic and medieval scholars. Repeated copying inevitably introduced errors, and the evolving use of language potentially distorted meaning.

Vesalius followed his father's interest in medicine, and in 1533 he moved to Paris to learn from some of the brightest educators of his day – Sylvius, Johann Günther, Jean Fernal – who were applying the principles of humanist scholarship to medical knowledge. War between France and the Holy Roman Empire forced Vesalius back home in 1536, but not before he had taken the highly unusual step of personally wielding the knife in his first dissection, undertaken in front of his fellow students.

His interest in anatomy continued in Leuven. Here he spent the night outside the city walls, with an anatomically inclined friend, to collect and smuggle in the bones of a gibbeted corpse. This escapade yielded a valuable, almost complete skeleton. Such touches of drama were not entirely gratuitous. Corpses were hard to obtain, centres of learning lacked their own skeletons, and Vesalius believed that the body was the essential source for anatomical knowledge. He completed his bachelor's thesis (published early in 1537) and continued his anatomical investigations before leaving for Italy later in the year.

Vesalius arrived in Padua in September 1537. On 5 December, he was awarded his doctorate. He had sufficiently impressed one of Europe's most renowned medical faculties to be offered a job as lecturer in surgery and anatomy. Vesalius began dissecting before his students the very next day.

Preparing the *Fabrica*

The University of Padua, under the protection of the Doge of Venice, was at the forefront of medical and surgical education. The medical faculty conducted at least one annual dissection – lasting three weeks during winter – attracting students eager for the new observational anatomy. The abdomen was opened first, to study its internal organs. The chest and its organs followed. The head and brain, and finally the limbs, were scrutinized. This time-honoured order maximized the longevity of each corpse as putrefaction set in.

Vesalius introduced two crucial pedagogic innovations at Padua; these were eventually writ large in the *Fabrica*. Unlike his contemporaries, Vesalius held the knife himself, acting as lecturer, demonstrator and dissector combined. He declined the lecturer's lectern, nor did he read from a book beside the body, but talked his way through what his hands revealed to the audience. He did not eschew all extant anatomical literature, but his humanist approach directed him back to the 2nd-century authority of the justly revered Galen of Pergamon, rather than commentaries on Galen, such as the popular 14th-century *Anathomia* of Mondino de Luzzi. Galen had instructed his readers to make a visual comparison between what was written and directly observed. Vesalius was already sufficiently experienced to realize that some of Galen's knowledge was derived from animals and therefore contained factual errors – issues to be discussed with the students. He continued to use animals, however, as necessary.

Among numerous other difficulties, dissecting region by region meant the study of body-wide systems such as the veins and arteries had to be reconstructed after the event (Vesalius, following Galen, has these as two separate systems based on the liver and heart). To help his students with such visualizations, Vesalius created true-to-life drawings to accompany his dissections. Subsequently reproduced as large pin-up sheets, the six illustrations (with some text) of the 1538 *Tabulae anatomicae sex* allowed nature and its literal representation to be studied side by side. Vesalius and the artist Jan van Calcar, a fellow Netherlander, drew the *Tabulae*'s illustrations. Calcar had studied with the great Italian painter Titian, at his studio in Venice. Padua's proximity to one of the world's leading artistic centres was extremely fortuitous.

The *Tabulae* was a great success. This can be measured by the swift and extensive plagiarism that followed. Rather more to Vesalius's taste was the additional supply of bodies (from executed criminals) offered to him in Padua. In January 1540, he was invited to share his new way of teaching anatomy at Bologna. In a showpiece before an audience of about 200 in the church of San Francesco, he clashed with the incumbent professor of anatomy Matteo Corti, who read from Mondino's *Anathomia* while Vesalius cut the flesh. The aging Corti derided such manual labour, but through it Vesalius publicly corrected Corti, Mondino, Galen and his own depiction in the *Tabulae*: the liver did not have five lobes.

Vesalius spent the next two years preparing the *Fabrica*. He read his Galen and as he worked at the anatomy table, there was an artist by his side. The drawings were transformed into astonishingly detailed woodcuts, often by pasting the drawing onto the wood, which was then cut through. Woodcuts had proved a wonderful innovation in printing. Functioning like the letter pieces of moveable type, they could be slotted into the page as desired. For important as his dense, classical Latin text was, the glory of the *Fabrica* lies in the pictures and the close association between word and image. The artist (possibly artists) from Titian's studio remains unidentified, but his contribution – under Vesalius's direction – was immense. The *Fabrica*'s frontispiece is a potent symbol of the new Vesalian science of anatomy. The sequences of skeletons and muscle men inevitably elicit gasps of admiration. Each of the fourteen muscle men appears in a continuous panorama and is posed in a mannered posture, as if, although flayed and revealed layer by layer, their bodies were still alive.

Vesalius oversaw the production process and left Padua for Basle to work with the humanist scholar and publisher Johannes Oporinus as the *Fabrica*'s pages were made up. Oporinus was renowned for the quality of his printing. His skill showed in the final product. The Fabrica was a triumph of Renaissance humanism. It was also an expensive, luxury item, not something a student could afford, nor what a working anatomist would want to consult with gory hands.

This niche in the market was met by the simultaneous publication of the much cheaper and briefer *Suorum de humani corporis fabrica librorum epitome*. Some of the *Epitome*'s loose sheets were designed to be cut up and the illustrations of organs placed on top of each other within a full-length body so a composite picture could be built up into a so-called flap-anatomy, popular at the time. A second, amended edition of the *Fabrica* appeared in 1555. Vesalius spent the rest of his life in the service of the *Fabrica*'s dedicatee, the Emperor Charles V, although he may have been contemplating a return to Padua before his death on the island of Zakynthos.

Afterlife

It is easy to read the *Fabrica* out of context; such was and remains its power. Earlier books on anatomy used few pictures, and these were mainly diagrammatic representations of the text: aides-memoire rather than visualizations of what one might see in the flesh. Vesalius was part of a Renaissance tradition of improving anatomical knowledge and depiction, but his work was a quantum leap forward. Besides the five-lobed liver, he corrected various other anatomical inaccuracies: humans do not share the rete mirabile (a complex of closely associated veins and arteries) with other vertebrates; the septum of the heart does not have pores to allow the blood to pass through.

Vesalius's illustrations were frequently copied and reused in subsequent texts over the next 100 years or so. True-to-life illustrations were eminently correctable in the face of new knowledge. Traditionalists criticized the twenty-eight-year-old Vesalius for criticizing Galen, but the spur to delve into the body to find that knowledge, rather than into a book, reverberated throughout Europe. Vesalius brought to anatomy the careful scrutiny of nature newly extended to other observational sciences such as botany and geography. The *Fabrica* helped to redefine the way we teach, learn and think about what lies beneath our skin for ever.

William Harvey

Experimental physician who discovered
the circulation of blood

(1578–1657)

Nature is herself to be addressed; the paths she shows us are
to be boldly trodden; for thus, and whilst we consult our
proper senses, from inferior advancing to superior levels,
shall we penetrate at length into the heart of her mystery.

WILLIAM HARVEY, *EXERCITATIONES*
DE GENERATIONE ANIMALIUM, 1651

It is remarkable to consider that there may be a single publication
that has determined the course of biomedical science and medicine
more than any other over the past millennium and a half. It is not an
overstatement to suggest that this book is William Harvey's *Exercitatio
anatomica de motu cordis et sanguinis in animalibus* (Anatomical study
of the motion of the heart and of the blood in animals), published
in 1628. This slender volume overturned the prevailing dogma of
physiology and medicine that had held sway in the western world
since the time of Galen of Pergamon.

Harvey was descended from a wealthy family of Folkestone. He
was the eldest of seven sons, five of whom became London mer-
chants. He obtained a bachelor of arts degree from the University
of Cambridge in 1597, before setting off on an extended study tour of
France, Germany and Italy. In 1602, he was awarded a doctorate in
medicine and philosophy from the University of Padua, where he was
exposed to the anatomical teaching of Fabricius ab Aquapendende

(also known as Girolamo Fabrizio), who had discovered the existence of valves in the veins. After graduating, he immediately returned to London, where in 1604 he was elected a fellow of the Royal College of Physicians.

Harvey's reputation was considerable during his lifetime. He was physician extraordinary to King James I and to his son Charles I; he had an extensive clinical practice and he was a physician at the oldest of the London hospitals, St Bartholomew's. He was a man of immense patience and persistence and a very careful scientist. It took him a quarter of a century to bring his studies to fruition. He once complained to a friend that as a result of his publication on the circulation of blood, his medical practice suffered, that he was being viewed as 'crack-brained', and that his colleagues envied his fame. He was obviously aware of his important place in the history of medicine and ensured that his reputation would be protected into the future. His bequest to the Royal College of Physicians and the gift of the house in which he was born and some adjoining lands to Caius College, Cambridge, where he had received his education, are probable testaments to this.

Blood and the heart

Galenic doctrine held that blood ebbs and flows to and from the lungs and the liver to the right side of the heart, and after passing through the ventricular septum, ebbs to and fro between the left ventricle and the arteries of the body. In the early 17th century, it was held that the heart is the source of heat and that the lungs serve to cool the blood. Diastole (expansion) of the heart supposedly brought blood and air together and the warmer, vitalized blood was then expelled into the systemic circulation. The fact that blood is darker in the veins and brighter in the arteries was attributed to the different functions of the two types of vessels in terms of their ability to nourish tissues and to retain vital spirits.

Harvey's research observations completely overturned these notions. He discovered that the left ventricle of the heart propelled the blood entering from the lungs in a continuous, unidirectional pattern into the major arteries and the tissues, and thence the blood returned via the veins to the right ventricle, which propelled it through the lungs. To arrive at this understanding, Harvey realized that the same amount of blood that leaves the veins must enter the arteries, to prevent one system from exceeding the other. For this to occur, the blood in the periphery must pass from arteries to veins, requiring a circular motion. The same principle had to apply to the circulation through the lungs: blood must flow from right to left ventricles via the lungs.

The observation that the heartbeat is simultaneous with the peripheral pulse had led to the misconception that the heart and the arteries expand and contract synchronously. Thus, when the heartbeat was palpable, the heart was thought to be in dilatation. By direct observation through the chest wall of animals, though, Harvey was able to show that this notion was simply wrong. The palpable heartbeat is due to contraction and rising up of the heart as it expels blood and is rotated to push up against the ribs. Thus, it was, that systole (contraction) and not diastole (expansion) of the cardiac ventricles coincided with the arterial pulsation.

In formulating his argument for the circulation of the blood in his *De motu cordis*, Harvey employed quantitative reasoning, which was quite a novel approach. He could not believe that ingested food alone could be the source of the large amount of blood constantly entering the heart via the veins. He also realized that the amount of blood flowing through the blood vessels must be far in excess of the amount required to nourish the various parts of the body. This simple reasoning implied to him that a fixed amount of blood must be moving around the body in a 'circular' fashion. This contribution was truly revolutionary and took many years to achieve wide acceptance.

But Harvey not only 'discovered'; he also 'created' an experimental method by which biological and medical experiments would be performed for centuries beyond his lifetime. He always started by asking questions (over twenty of them in the first chapter of the *De motu cordis*), some of which politely ridicule extant views and for which the answers are all too obvious. The questions form the basis for the experiments that follow. Central to his method was his use of vivisection – observations on live animals – an experimental technique that unlocked all doors for him. An observation at a single point in time (that is, a dissection of a dead animal) was not sufficient to answer questions of function. Continuous, serial observations in live animals were required. Ligatures, excisions and exposures of body parts were his means for manipulating and exposing normal physiology.

A curious mind

From the earliest days of his professional career, Harvey understood that every new species that he studied added a new insight. The range of animal species that he studied was very large. It was when Harvey moved into embryogenesis in animals, later in life, that his intensely curious mind was at its most extended. Despite the objectivity that he demanded of himself, he could not but retain a fascination with creation and what lay behind it. He was drawn to the study of embryology and to the earliest stages of development. What comes first? What follows? Such thoughts are evident in the last book that Harvey published, in 1651, *Exercitationes de generatione animalium* (*On the generation of animals*).

Harvey was a revolutionary figure who brought to medicine an immense curiosity together with a rigorous and disciplined thought process in conducting biological experiments. The result was a fundamentally new understanding of how the human body works. He had shown his disciples the way to ask the right questions, and how best to try and answer them.

Louis Pasteur

Revolutions in the treatment of disease

(1822–1895)

In the fields of observation, chance favours
only prepared minds.

LOUIS PASTEUR, IN A SPEECH AT THE UNIVERSITY OF LILLE, 1854

By the time of his death in 1895, Louis Pasteur was a French national hero and an international celebrity. He was best known to the public for his work on the prevention and treatment of infectious diseases in his later years, but to the scientific community he was the man who had created the field of stereochemistry, revealed the biological nature of fermentation, defeated the doctrine of spontaneous generation, helped establish germ theories of disease, and demonstrated across many fields the economic and social benefits of experimental laboratory research. Pasteur's standing was due in part to astute self-promotion, but came mainly from the breadth of his achievements in the theory and practice of microbiology. Yet, as Gerald Geison wrote: 'Although often bold and imaginative, his work was characterized mainly by clearheadedness, extraordinary experimental skills, and tenacity – almost obstinacy – of purpose.' At all levels, from his choice of research topics to serendipitous results from single experiments, Pasteur enjoyed good fortune. It was because he exemplified his own much-quoted maxim about the 'prepared mind', as he had the knowledge, insight and creativity to exploit his opportunities.

Pasteur was born in Dole in eastern France, the son of a tanner. He attended school in Arbois and Besançon and was academically strong enough to be recommended for the entrance examination of the prestigious École Normale Supérieure in Paris. He failed the exam in 1842, but passed a year later. He chose the physical sciences, and after doing well in his first degree, went on to pursue a twin-track doctorate in physics and chemistry, focusing on the new field of crystallography.

He studied the relationship between the chemical formula and crystalline forms of sodium tartrate. Scientists were interested in different chemicals that had very similar crystal structures – or isomorphism. The salts of tartaric acid were of particular interest because they showed dimorphism – that is, two forms of the same chemical. Pasteur followed a number of research paths to explore this phenomenon, and microscopy was crucial in revealing that the two forms were mirror-images of each other. Luck played a part: crystallization is highly sensitive to temperature and Pasteur worked at the optimum time of the year. Also, sodium tartrate displays asymmetry more clearly than almost any other salt. He found that crystals from natural sources had polarizing properties, while those synthesized in the laboratory did not. Close and painstaking observation revealed that natural crystals were all right-handed, while synthetic ones had equal quantities of right- and left-handed forms, which meant that their polarizing properties cancelled each other out. Five features of this investigation began to define Pasteur's scientific style: his skill and tenacity as an experimentalist, his use of microscopy, his interest in the uniqueness of the chemistry of life, his ability to make the most of his luck, and the high impact of his results.

In 1849, Pasteur moved to the University of Strasbourg as professor of chemistry, continuing to work on asymmetry and enjoying his growing scientific reputation. His personal life changed when he married Marie Laurent, the daughter of the rector of the university, who gave devoted and practical support to his career. After six years he moved to University of Lille as dean of the new Faculty of Science, embracing the university's mission of linking teaching and research with supporting local industries through the application of science. He taught bleaching, refining and brewing, but his research continued to be on asymmetric compounds and their optical activity.

Fermentation

Pasteur's fascination with the chemistry of living organisms led to investigations of fermentation and particularly the role of yeast in the production of alcohol. In 1857 he published on lactic acid, which was a common by-product of abnormal fermentation, and on amyl alcohol. Pasteur maintained that the asymmetric optical properties of amyl alcohol came from the process of fermentation, which confirmed his view that it was due to living organisms. This went against accepted ideas of fermentation as a chemical process. In 1860, Pasteur, who was back at the École Normale as director of scientific studies, published a major study on fermentation that was decisive for the biological explanation. It was perhaps ironic that someone trained in chemistry and physics was supporting vitalism, a view that stressed the uniqueness of life and claimed its phenomena could not be reduced to material forces. Pasteur had turned his microscope from crystal structures to looking at grapes fermenting and milk souring, observing that yeast and other 'ferments', previously assumed to be large molecules, actually changed shape during the process, further confirming that yeast was composed of living cells or their germs.

During his studies of fermentation, Pasteur was drawn into a famous controversy with Félix-Archimède Pouchet over the spontaneous generation of life. Pouchet became vocal in his advocacy of spontaneous generation at the end of the 1850s. Pasteur first spoke out against it in February 1860, and the following year published a prize-winning essay arguing that life always sprang from prior life. His objects of study were fermentation and putrefaction in infusions of natural products, which he maintained were always due to contamination by living ferments; Pouchet argued that they could arise spontaneously without contamination. The two scientists joined in a scientific duel, exchanging experimental results and polemics, in which fine matters of technique in sterilization were mixed with reflections on the religious implications of whether life was constantly being created. Pasteur sided with the conventional view that life had been created in the distant past by God's actions and could not arise simply from physical forces. The contest was settled in Pasteur's favour and against spontaneous generation, not only through the emerging consensus of the scientific community, but also and unusually, in judgments handed down by committees of the French Academy of Sciences.

Germs and disease

The controversy drew Pasteur into new studies of disease in animals and humans. Doctors had long viewed the development of fevers and septic infections as analogous to fermentation and putrefaction; knowing these processes to involve living organisms or their germs posed new questions. The link was, of course, speculative, and this was nicely captured in the phrase 'germ theory of disease'. 'Germ' carried the sense that the organisms were protean, widely spread in the environment, especially in the air, and potentially powerful through multiplication; 'theory' implied that the link to disease causation had still to be proved. Pasteur pursued the practical application of his germ theories of fermentation and putrefaction, finding that heating wine to 50 °C killed the yeast cells and prevented deterioration. The same method, when applied to stop milk souring, is still known as pasteurization. The most famous medical use of Pasteur's germ theories was by the British surgeon Joseph Lister, who assumed that septic infection of wounds was due to contamination by putrefying germs and developed methods of antisepsis. Lister became a champion of the wider application of germ theories to all infectious and contagious diseases, always paying homage to Pasteur in his campaign.

Pasteur's reputation for applying science to practical problems led the French government to ask him to head a team to investigate disease in the silk industry in 1865. After three years of investigation, the disease was linked to a parasite, and practices were recommended for keeping silkworms germ-free and healthy. Pasteur's success gave a higher profile to germ theories of disease, which medical investigators across the world increasingly adopted to frame their research and practice. But during this work, Pasteur had his first stroke and developed a weakness on his left side that persisted for the rest of his life, though it did not affect the pace or ambition of his work. Indeed, he entered perhaps the most productive period of his career.

Pasteur's first study of an infectious disease was with anthrax, which was mainly a problem for the French livestock industry, but could also affect humans. Its bacterial cause was established by Robert Koch in 1876. Although Pasteur disputed aspects of Koch's work, he is best known for his manufacture of a vaccine. Following the principle of vaccination against smallpox, that a mild infection can protect against a more serious one, he set out to reduce the virulence of anthrax bacteria by exposure to air. He had laboratory success and moved to a field trial at Pouilly-le-Fort near Paris in 1881. Twenty-five sheep were vaccinated and an equal number left as controls. Two weeks later, all were inoculated with anthrax germs. Almost all of the vaccinated sheep survived, whereas almost all of the unvaccinated died. Besides promising immediate benefits for French farmers, Pasteur's work had shown that vaccination might be applied to many if not all infectious diseases. He was lionized at the International Medical Congress in 1881, and enjoyed the generous support from the French state.

New vaccines

Yet greater things were to come. His next project was a vaccine to protect against rabies, which though relatively rare, excited great public anxiety because of its unpredictability and the fact that symptoms, once started, inevitably led to the worst of all deaths. First, Pasteur and his growing number of assistants produced the disease in laboratory dogs and rabbits under controlled conditions. Early trials on dogs were successful and were extended to humans, in part due to public pressure and in part to growing confidence in the new vaccine. However, the rabies vaccine was used to treat those who might already be infected, not for prevention. The idea was to take advantage of its long incubation period to build up immunity. The first public trial was with a boy, Joseph Meister, bitten by a rabid dog in eastern France and brought to Paris by his parents, who had read of Pasteur's possibly life-saving cure. The boy survived the treatment. Then the vaccine was tried successfully on another boy. After a public announcement was made in October 1885, rabid dog-bite victims from across France and Europe, and soon the world, flocked to Paris to receive a treatment that was free to everyone.

The new popular press made Pasteur's rabies cure front-page news, hailing him as a great scientist and humanitarian whose work promised to deliver humankind from the spectre of infectious diseases. More awards and rewards followed, but the most significant was a public subscription to found an institution to develop further vaccines and other life-saving innovations. Money flowed in and the Pasteur Institute was inaugurated in November 1888. By this time, the great man's health had deteriorated, and while he was a presence in the laboratory and clinic, he was no longer active in research. Upon his death in 1895, he was given a major public funeral with the offer of interment in the Panthéon with other French heroes. But Pasteur and his family had already made their plans. He was laid to rest in the crypt of his institute, a fitting place for a scientist whose work had redefined the place of laboratory research in science and public affairs.

Francis Galton
Explorer, statistician, psychologist
and inventor of eugenics
(1822–1911)

I take Eugenics very seriously; feeling that its principles
ought to become one of the dominant motives in a civilised
nation, much as if they were one of its religious tenets.

FRANCIS GALTON, *MEMORIES OF MY LIFE*, 1909

Francis Galton, often remembered as the father of the eugenics movement, was a man of many interests who made important contributions
in areas as diverse as African exploration, psychology, statistics and
fingerprinting. Francis was the last of Tertius and Violetta Galton's
nine children. He shared a common grandfather, Erasmus Darwin,
with his older cousin Charles Darwin. Like Darwin, Galton first
appeared destined for medicine. He began learning his trade at the
General Hospital in Birmingham, but transferred to King's College
in London, in 1839. Darwin, fresh from his *Beagle* voyage and newly
married to Emma Wedgwood, was residing nearby. Darwin, who
hated his medical experience at Edinburgh, apparently convinced
Galton to abandon medicine for Cambridge, where he had studied.
In October 1840, Galton went up to Trinity College, hoping for an
honours degree in mathematics, but graduated with only an ordinary degree.

After drifting for six years, Galton organized an expedition that
visited unexplored parts of Namibia, discovering a new tribe, the
Ovambo, and making precise measurements of latitude, longitude

and temperature. He returned to England in early 1852 and was awarded the Founder's Medal of the Royal Geographical Society. In 1853, he married Louisa Butler, the daughter of George Butler, dean of Peterborough Cathedral, and published his first book, *Tropical South Africa*. Two years later, he published *Art of Travel*, an enormously successful guide. He also became interested in making retrospective weather maps, in the process discovering the high-pressure system in which winds circulate clockwise.

Nature versus nurture?

These various forays defined the first part of Galton's career. The second began with the publication of Darwin's *On the Origin of Species* in 1859. Darwin used examples of artificial selection – fancy pigeons, for instance – to illustrate how natural selection might work. Galton realized that if selection worked for pigeons, it could also work for people. Perhaps the human race could be improved through selective breeding.

In 1865, Galton published an article in *MacMillan's Magazine*, one of numerous high-quality Victorian periodicals, entitled 'Hereditary talent and character'. In this article and his subsequent 1869 book *Hereditary Genius*, he examined the close relatives of famous men. He reasoned that if 'talent and character' were hereditary, then those male relatives most closely related to the eminent man were more likely to be eminent than more distant relatives. Galton concluded that this was indeed the case, dismissing the possibility of 'jobbery' – that is, that the eminent man might have secured a position of distinction for his son. *Hereditary Genius* has been described as the first example of historiometry, the historical study of human progress or individual personal characteristics. Galton was also the first to use the phrase 'nature versus nurture' in this context. He even devised a questionnaire that he sent out to 190 fellows of the Royal Society in order to provide a firm evidential foundation to the issue.

He tabulated characteristics of the men's families and attempted to discover whether their interest in science was 'innate' or due to the encouragements of others. The studies were published as a book, *English Men of Science: Their Nature and Nurture*, in 1874.

In 1875, Darwin published a second edition of *The Variation of Animals and Plants under Domestication*. What intrigued Galton was Darwin's chapter entitled 'Provisional hypothesis of pangenesis'. Darwin needed a source of variation for natural selection to work on. He postulated that particles he called gemmules were gathered from throughout the body 'to constitute the sexual elements, and their development in the next generation forms a new being'. Darwin imagined two mechanisms for generating variation. First, the reproductive organs might undergo an injury that prevented the gemmules from aggregating properly. Second, the gemmules could be modified 'by direct action of changed conditions'. These modified gemmules were then transmitted to the offspring and, after several generations, the alteration became heritable.

Galton was intrigued by Darwin's hypothesis, although he did not like his idea that gemmules could be modified by altered environmental conditions. So Galton went on to construct his own theory of inheritance, a version of the germ-line theory of the German evolutionary biologist August Weismann – according to which inheritance can take place only via the germ cells (that is, the egg and sperm cells), meaning that acquired characteristics cannot be passed down between generations – a point that Weismann himself acknowledged in a letter to Galton in 1889.

But Galton was less a theorist than a practical scientist. He wanted numbers to analyse, eventually ones relating to human traits. With the advice of Darwin and the botanist Joseph Hooker, he decided to make measurements on sweet-pea seeds. One reason why he chose sweet peas was that they usually did not cross-fertilize. Galton found that seed sizes were distributed normally among the progeny as they had been among the parents. But he also found that the mean size of seeds from, for example, a parent selected for large seeds reverted to the mean. This was true of small seeds as well. When Galton plotted the average diameter of the parental seeds on the X-axis and of the progeny seeds on the Y-axis, he got a straight line. It was the first regression line, now one of the central principles of statistics. From it, he calculated the first regression coefficient – or coefficient of reversion, as he called it.

In 1884, Galton established an anthropometric laboratory at the International Health Exhibition in South Kensington, London. Visitors were given a card on which a variety of measurements were recorded. Galton also managed to collect partial pedigree data. From these, he showed that regression to the mean applied to height in people, too. He also discovered that when he plotted a measurement like forearm length against height, the measurements were correlated and so the first correlation coefficients were born, another milestone in the history of statistical analysis.

At the close of the International Health Exhibition in 1885, Galton moved his anthropometric laboratory to the Science Galleries of the South Kensington Museum (today the Victoria & Albert Museum). He now included a place for a thumbprint on his questionnaire as he had become interested in fingerprint patterns. His friend Sir William Herschel of the Bengal Civil Service had made the key observation that fingerprint patterns were stable over time. In the 1890s, Galton published two books on fingerprints and played a central role in establishing fingerprinting as a means of personal identification.

Inheritance and eugenics

In 1889, he published his most important book, *Natural Inheritance*. This volume inspired his three main disciples: Karl Pearson, W. F. R. Weldon and William Bateson. Galton's chapter on the normal distribution and the continuous variation of characters was what interested

Pearson and Weldon, but Bateson was intrigued by something else. Galton had been wrestling with a problem. How could natural selection work by small incremental steps if it was to be thwarted by regression to the mean? To finesse the problem, Galton proposed his 'organic stability' hypothesis for establishing new variants that could not regress to the mean. For Bateson, these discontinuous variants were exciting. He had collected many examples of discontinuous variation and published them in his 1894 *Material for the Study of Evolution*. The upshot of all this was that Bateson was mentally prepared for the rediscovery of Gregor Mendel's principles in 1900. Mendel's laws described the segregation and assortment of discrete characters of the kind Bateson was interested in – yellow versus green pea seeds, for example.

Pearson and Weldon, in contrast, were firm supporters of a new model of Galton's proposed in 1898 called the Law of Ancestral Inheritance. The ancestral law really applies to the whole genome, since it contemplates a continuous series with parents contributing one-half (0.5) to the heritage of their offspring, grandparents one-quarter $(0.5)^2$, great grandparents one-eighth, etc. The whole series $(0.5) + (0.5)^2 + (0.5)^3 \ldots$ adds up to 1. Pearson and Weldon tried to make Galton's Ancestral Law fit discrete characters, but Bateson foiled them at every turn by demonstrating that Mendel's principles fit the data much better.

Galton was also interested in measuring human intelligence. His anthropometric laboratory data provided a first approximation. But then he had a better idea. He knew that twins could be of two kinds that we now call identical and non-identical. He published his findings in an 1875 article in *Fraser's Magazine*. He found that identical twins, besides being physically similar, often shared behavioural quirks. He could not quantify intelligence because the IQ test had yet to be invented, but his results suggested a marked heritable component to behaviour and by extension to intelligence.

Galton defined eugenics in a footnote in his 1883 book *Inquiries into Human Faculty and its Development*. Eugenics, he wrote, deals with 'questions bearing on what is termed in Greek *eugenes*, namely, good in stock, hereditarily endowed with noble qualities'. In speeches and papers, he kept plugging eugenics and it became popular early in the 20th century. But it was not Galton's idea of improvement of the best stock that caught on (positive eugenics), but elimination of stocks perceived to be inferior (negative eugenics). The momentum began to build at the First International Congress held in London in 1912, the year after Galton died. There were many unfortunate and unintended outcomes, in particular the forced sterilization of women perceived as inferior or mentally defective in the United States, the Scandinavian countries and, most notably, Nazi Germany.

So Galton leaves a mixed legacy. He made important contributions in areas as diverse as African exploration, travel writing, statistics and fingerprinting, but he also founded eugenics, with all the horrific consequences of that movement.

Sigmund Freud

Theorist of the unconscious and the founder of psychoanalysis

(1856–1939)

The view that the psychical is unconscious in itself enabled psychology to take its place as a natural science like any other. The processes with which it is concerned are just as unknowable as those dealt with by other sciences, by chemistry and physics, for example; but it is possible to establish the laws which they obey and to follow their mutual relations and interdependencies unbroken over long stretches – in short, to arrive at what is described as an 'understanding' of the field of natural phenomena in question.

SIGMUND FREUD, *AN OUTLINE OF PSYCHOANALYSIS*, 1940

Sigmund Freud was born to a relatively poor Jewish family in Freiberg, now Pᵛríbor in the Czech Republic. When he was four the family migrated to Vienna, where Freud later enrolled in the university medical school, which was undergoing a golden age. There he specialized in neurology, studying in the 1870s under Ernst von Brücke, a leading member of the experimental Helmholtz school of medicine, before travelling to Paris in 1885 to study under Jean-Martin Charcot, the first university professor of neurology and champion of clinical observation. Though brief, the visit to Charcot was a turning point because it made Freud realize that some neurological disorders (such as hysteria) were best understood psychologically. On his return to Vienna in 1886 he set up in clinical practice, eventually specializing in such disorders. He remained there until the Nazi take-over in 1938, when he fled to London (where he had been elected a corresponding member of the Royal Society in 1936), the year before he died.

Subjecting the subject to scrutiny

What distinguished Freud from other scientists was his subject matter. All scientists study parts of nature, such as stars, mountains, birds, bees, molecules or atoms. These things, no matter how large or small, are *objects*. Scientists aspire to describe them objectively (as they really are), not merely as they appear to us subjectively. Subjectivity is considered a major source of error in science. Freud, by contrast, made the observing subject his object of study – the very thing that other sciences had sought to exclude. This was bound to get him into trouble.

It is difficult to deny the fact that subjectivity is part of nature. It exists. As René Descartes famously concluded, subjectivity is the part of nature about which we can be the most certain: 'I think, therefore I am.' But things like thoughts (and feelings) do not exist in the objective world; they exist only within ourselves. This led Descartes to another famous conclusion. Nature, he said, seems to be made from two kinds of stuff: physical and psychological. This conclusion made it easy for scientists to exclude the psychological part of nature from their studies. But it did not do away with it; psychological stuff (like thoughts and feelings) still existed. It existed, but outside of science. This was bound to get science into trouble.

How do psychological causes have physical effects? How does a thought – 'I shall move my finger' – cause the finger actually to move? This is the infamous 'mind–body problem', which has bedevilled philosophers ever since Descartes. For scientists, his conclusion must be wrong; things with mass and energy cannot be affected by things without mass and energy. The second law of thermodynamics (the law of the conservation of energy), which is fundamental to our understanding of physics, contradicts this possibility outright. It implies that Descartes's philosophy has to be abandoned, and so in time it was.

There were two mainstream alternatives to Freud's approach. The first was the one that Freud himself had taken before he developed psychoanalysis. The second came later, in opposition to psychoanalysis. The first approach was to study not the mind itself but rather its physical 'scene of action' – the brain – and attempt to infer the laws of the mind from the functions of the brain. This is what Freud did until 1895. Many, if not most, scientists who adopted this method went so far as to assert not only that it was more scientific (more objective) to study the physical correlates of subjectivity than subjectivity itself, but that subjectivity *does not really exist*. It consists in mere appearances, ultimately reducible to physical things. The problem with this sleight of hand is that it takes us back to our starting point, and once again excludes subjectivity from science. This is because no one is able to explain how the subjective appearances are reducible to physical things, or, to put it another way, how the physical things cause the appearances. Freud accordingly abandoned this approach, as he said, 'in 1895 or 1900 or somewhere in between' – before publishing *The Interpretation of Dreams* in 1899.

The other mainstream alternative was 'behaviourism'. This approach, which came to prominence in the 1920s, did not study the mind directly either. Rather it studied the observable inputs and outputs of the mind – its *responses to stimuli*. From these observable events, behaviourists inferred the laws that generated the responses. These were the laws of the mind. Although their methodological assumptions did not seem to demand it, most behaviourists went further, and claimed also that the mind itself (subjectivity) did not really exist. They reduced the laws of the mind to a mechanism called 'learning'. It is not difficult to guess why they did this: the intrinsic property of the mind – subjectivity – is an embarrassment to science. It cannot be controlled experimentally. It is not in the nature of the mind to behave like an object. The mind is not an object. The mainstream approach today – cognitive neuroscience (descended from both neuroscience and behaviourism) – still largely disregards this fact.

Bridging the mind–body divide

What, then, was Freud's approach? He took the fact of subjectivity as his starting point (calling it 'a fact without parallel'). He then closely observed thousands of instances of subjective experience, in a standardized setting. On this basis he tried to infer the laws that underpinned experience. Freud was well aware that, in doing this, he was not conducting normal science: 'It still strikes me myself as strange that the case histories I write should read like short stories and that, as one might say, they lack the serious stamp of science. I must console myself with the reflection that the nature of the subject is evidently responsible for this, rather than any preference of my own.'

Freud is therefore largely remembered through his case histories. The first and perhaps most famous was that of 'Anna O' (in fact she was the patient of a colleague, Josef Breuer), who spontaneously observed that her symptoms recovered when she talked about a psychological trauma that had triggered them. This was the origin of the 'talking cure'. Freud went on to report similar observations in numerous cases of hysteria (such as 'Dora') and other neuroses (such as the 'Rat Man', 'Little Hans' and the 'Wolf Man'). His principal discovery along the way was that the events that trigger neurotic symptoms can be understood only if they are traced back further, to the patient's earliest attachments, which are ultimately reducible to the instinctual constitution of our species, expressed in the form of powerful sexual and aggressive feelings towards care-givers (the infamous Oedipus complex). Freud concluded that the basic mechanism of neurosis was unsuccessful 'repression' of these instinctual drives. He then contrasted this with the mechanism of other mental illnesses, such as psychosis (in the case of Judge Schreber), which entails an unsuccessful attempt to disavow not the instincts but the things in the outside world that frustrate them.

Freud called the laws that he inferred in this way 'metapsychology'. With this concept, he tried to resolve the mind–body problem (he aimed 'to transform metaphysics into metapsychology'). For Freud, the laws that underpin subjective mental life must be similar in kind to those that underpin physical life. This was demanded by his teachers from the Helmholtz school, who swore this solemn oath: 'No other forces than the common physical-chemical ones are active within the organism. In those cases which cannot at the time be explained by these forces one has either to find the specific way or form of their action by means of the physical-mathematical method, or to assume new forces equal in dignity to the chemical-physical forces inherent in matter, reducible to the forces of attraction and repulsion.'

The psychological forces that Freud assumed (libidinal drive, repression, disavowal, etc.) provided him with a language to describe the functional organization of the mind, which he knew must parallel the functional organization of the brain. Just as Freud had initially attempted to infer the laws of the mind from the functions of the brain, he later tried to infer the laws of the brain from the functions of the mind. He made this methodological reversal for purely expedient reasons: scientific tools were then lacking for studying the functions of the brain.

The laws that Freud inferred from his observations were functional (abstract) laws, like those of modern cognitive psychology, not physiological (concrete) ones. It is not generally known that Freud pioneered this 'functionalist' approach. He used it to accommodate the fact that although mind and brain are different observational starting points, they are ultimately the same 'thing' and therefore must share a common underlying organization. (He called this abstracted thing the 'mental apparatus'.) That is why Freud insisted on the strange-sounding phrase: 'the psychical is unconscious in itself'. The concept of an unconscious mental apparatus provided him with the 'long-sought missing link' between mind and body. Today Freud's scientific heirs are reviewing and revising his conclusions with the tools of modern neuroscience (such as functional brain imaging), while simultaneously trying to use them to correct the mistakes of those neuroscientists who still disregard the intrinsic properties of the mind, and thereby overlook some fundamental facts about how it works.

Alan Turing

The father of computer science and artificial intelligence

(1912–1954)

*Mathematical reasoning may be regarded rather
schematically as the exercise of a combination of two
facilities, which we may call intuition and ingenuity.
The activity of the intuition consists in making spontaneous
judgements which are not the result of conscious trains
of reasoning.... The exercise of ingenuity in mathematics
consists in aiding the intuition through suitable
arrangements of propositions, and perhaps
geometrical figures or drawings.*

ALAN TURING, *THE PURPOSE OF ORDINAL LOGICS*, 1938

Alan Turing has a rightful claim to be called the father of modern computing. In the years before World War II, he laid the theoretical groundwork for a universal machine that established the model for a computer in its most general form. His seminal 1936 paper, '*On computable numbers with an application to the Entscheidungsproblem*', with hindsight, foretold the capabilities of the modern computer. During the war, he was instrumental in developing and influencing computing devices that were said to have shortened the war by up to two years by decoding encrypted enemy messages thought to be unbreakable. Unlike some theoreticians, he was willing to be involved with practical aspects, and was as happy to wield a soldering iron as he was to wrestle with a mathematical problem, normally coming at it from a unique angle.

A precocious student

Alan Mathison Turing was born on 23 June 1912, the son of civil servant posted to India. His parents returned to Britain for his birth in Maida Vale, London. At the age of fourteen, he was sent to Sherborne School in Dorset, southern England, a traditional British public school. His interest in science was noted by his school teachers, but not particularly encouraged at such a conservative establishment. He was able to solve advanced problems from first principles, without having been taught calculus, for example. At sixteen, he encountered and understood the work of Albert Einstein. While at school, Turing formed a close friendship with a fellow student who died during his last term at Sherborne. This had a traumatic effect on him and any religious leanings he had were destroyed. He eventually became an atheist.

Turing went on to study mathematics at King's College, Cambridge, from 1931 to 1934, graduating with a first-class degree. He was subsequently elected a fellow of the college. In 1936, he submitted his groundbreaking paper on computable numbers that was to form the cornerstone of the rest of his career. This presented the concept of computing machines and in particular described a universal machine capable of computing a wide class of numbers. Turing's notion of universality is what is thought of as programmability in computers today. As he stated in a paper published in 1939, in the context of a human computer, 'a man provided with paper, pencil, and rubber, and subject to strict discipline, is in effect a universal machine'.

Turing's efforts developed the 1931 research of the German mathematician Kurt Gödel and led to the use of the term 'Turing machine' for his universal machine. He demonstrated that any mathematical calculation that can be represented as an algorithm can be performed by such a device. The *Entscheidungsproblem* addressed in his 1936 paper was a mathematical challenge posed by the German mathematician David Hilbert in 1928 as to whether there is always an algorithm that can determine the truth or falsity of a mathematical statement. Turing also demonstrated the insolubility of the problem by showing that it is impossible to decide algorithmically whether a given Turing machine is satisfactory. This is now known as the halting problem and the issue was a vexing one to mathematicians. Turing machines remain an important concept in the theory of computation to this day. Although an abstract concept – Turing had no intention of building one – it was, in fact, buildable and anticipated many of the features and processes of contemporary computers, such as input, output, memory and coded programs.

From 1936 to 1938, Turing studied under the American mathematician Alonzo Church at Princeton University, obtaining his doctorate in a remarkably short period. Earlier, they had independently developed the Church-Turing thesis, characterizing the nature of computation, stating that every effectively calculable function produced through any means is also computable using a Turing machine. Although the thesis cannot be proved, it is almost universally accepted by mathematicians and theoretical computer scientists. Turing returned to Cambridge and attended lectures by the philosopher Ludwig Wittgenstein about mathematical foundations. Wittgenstein argued that mathematics invented rather than discovered truth, but Turing disagreed.

Breaking the code

War brought about a radical but fortuitous change of direction in Turing's career. His unique mathematical abilities had been recognized during his time at Cambridge and he was invited to join Bletchley Park, the secret centre of Britain's efforts to break German codes. He was recruited after working part-time for the Government Code and Cypher School (now known as the Government Communications Headquarters or GCHQ). Code decryption was laborious by hand and difficult to achieve in the short amount of time available to a nation facing an impending enemy. Turing recognized that machines, together with great human ingenuity, could tackle the problem far more quickly and reliably. Before the war, he had already contributed ideas on breaking the Enigma machine used by the Germans for encrypting messages. It meant that within weeks of joining he had devised a machine that could be used to help decode Enigma. His solution was named the bombe after an earlier, less efficient Polish-designed device, the bomba. The bombe worked by taking a piece of likely plain text from the original message (known as a 'crib') and passing it through combinations of the Enigma's rotors and plugboard settings. Most possible settings would quickly produce contradictions, allowing them to be eliminated, leaving only a few combinations to be investigated in greater depth. The machine effectively undertook a mathematical proof mechanically, far more efficiently than a human, or even a team of humans, could ever do.

Turing chose to work on naval Enigma decryption because, as he said, 'no one else was doing anything about it and I could have it to myself'. This was typical of Turing, although he also collaborated well with others. He devel- oped a number of novel decryption methods during his time at Bletchley Park, often giving them playful slang terms. One of these, developed in 1942, was 'Turingery' or 'Turingismus', a hand technique for finding patterns in the wheel cams of the Lorenz cipher machine, another encryption device used by the Germans to encipher high-level strategic commands that had been discovered by the British. It was especially useful because the information remained valid for a significant period.

Some of Turing's eccentricities were quite clearly evident to his colleagues at Bletchley. He chained his cup to his radiator in his office so it would not be lost or stolen. He was also known to wear his gas mask while cycling to work, not because of fear of being gassed, but to avoid hay fever. Nonetheless, his influence on work at Bletchley Park helped in the development of the world's first programmable digital electronic computer there, the Colossus. Turing was awarded the OBE in 1945 for his war work, but his contribution remained secret for many years.

Testing times

After the war, Turing took up a post at the National Physical Laboratory in Teddington, west of London. There he worked on the design of the Automatic Computing Engine (ACE), an early computer. Unfortunately delays meant that even the cut-down Pilot ACE was not built until after he had left the laboratory in order to return to Cambridge for a sabbatical year.

In 1948, he joined the mathematics depart- ment at the University of Manchester. He was appointed the deputy director of the computing laboratory at the university, working on software for the Manchester Mark 1, an early stored-program computer. Turing continued to consider more theoretical and abstract ideas, including the concept now known as artificial intelligence, in which he explored whether a machine can think. He devised the Turing test as a possible demonstration of machine intelligence: to pass the test, a computing machine must appear human when interacting with a person, in such a way that it is indistinguishable from a real human being. Such a feat has still not been achieved, although it is deemed to be a viable aim and remains relevant to this day. In fact, there is now a whole raft of variants of the test in use.

Turing was a homosexual at the time when homosexuality was illegal in the United Kingdom. He was charged with gross indecency in 1952 and lost his security clearance as a result. Instead of imprisonment, he was forced to take female hormones for a year in an attempt to 'cure' him. From this point on, he worked in the interdisciplinary area of mathematical biology and specifically morphogenesis, the process that allows organisms to generate their shape. Much of this work was not published until his collected papers appeared in 1992.

On 8 June 1954, Turing was found dead at home by his cleaner. The cause of death was cyanide poisoning, believed to be via a half-eaten apple found by his bed, but this was never tested. It was determined that he had committed suicide, although it is possible that his death was an accident.

Turing had been made a fellow of the Royal Society in 1951, a year before his fall from grace, but the real recognition of his contribution came long after he died. There is now a memorial statue in Manchester and another at Bletchley Park. There are commemorative plaques marking his London birthplace and the home where he died in Wilmslow, Cheshire. In 2009, there was even an apology from the British government for his official treatment in the years leading up to his death. Perhaps most fittingly, the nearest equivalent to the Nobel Prize given annually to an outstanding computer scientist is known as the A. M. Turing Award. Despite his untimely death at only forty-one, Turing's influence will live on in the field of computing for the foreseeable future.

John von Neumann
Mathematician and designer
of the electronic computer
(1903–1957)

*There was something endearing and personal about
Johnny von Neumann. He was the cleverest man
I ever knew, without exception.*

JACOB BRONOWSKI, *THE ASCENT OF MAN*, 1973

John von Neumann was born in Budapest, the eldest of three sons of a prosperous and cultured Jewish banking family. He was educated by a private tutor until the age of ten, and then attended the Lutheran Gymnasium in the Hungarian capital. His remarkable talents were evident from an early age – he had an almost photographic memory and a startling facility for performing rapid mental arithmetic. At the age of eighteen, he enrolled for a mathematics degree at the University of Budapest, although he spent much of his time in Berlin getting to know Europe's scientific elite. He subsequently began doctoral studies at the University of Budapest, but also simultaneously enrolled at the Eidgenössische Technische Hochschule (ETH) in Zurich, to study chemical engineering at the behest of his father who wanted him to have a practical education. He obtained a degree in chemical engineering from ETH in 1925 and a doctorate in mathematics from the University of Budapest in 1926.

In 1926, he became a Rockefeller Fellow at the University of Göttingen in Germany, and the following year was appointed as a *privatdozent* (lecturer) at the University of Berlin, the youngest in the university's history. His researches in the 1920s were wide-ranging, and included mathematical logic, set theory, operator theory and quantum mechanics. In 1930 he became a visiting lecturer at Princeton University and for some years divided his time between there and Berlin. He was, however, anxious to secure a permanent position in the United States because of the deteriorating political situation in Europe. This opportunity came in 1933, when he was appointed one of the first four founding professors of the newly formed Institute for Advanced Study in Princeton (Einstein was another). He became a naturalized citizen of the United States in 1937. At the Institute, von Neumann maintained a prodigious output in pure and applied mathematics, as well as developing the theory of games. His 1944 book *Theory of Games and Economic Behaviour*, co-authored with Oskar Morganstern,was a landmark in mathematical economics.

Wartime calculations and the first electronic computer

Von Neumann had an engaging personality, was highly sociable, and politically astute. When the United States entered World War II following the attack on Pearl Harbor in December 1941, his bonhomie, legendary quickness of mind and ability to cut through complex mathematical problems created a great demand for his services as a consultant. By 1943 all his attention was directed to war work, particularly to problems of numerical computation. Most importantly, he became a consultant to the Manhattan Project in Los Alamos. There he advised on implosion techniques for detonating the fissile material at the core of the atomic bomb. This involved the numerical solution of complex systems mathematical equations and caused him to seek out the most advanced computing machinery available.

Von Neumann was also a consultant to the US army's Ballistics Research Laboratory based at the Aberdeen Proving Ground, Maryland. One of the laboratory's principal tasks was the production of ballistics tables, and it had funded a project at the nearby University of

Pennsylvania's Moore School of Electrical Engineering to construct an electronic computer, the ENIAC (Electronic Numerical Integrator and Computer). Because of technical and design limitations, the ENIAC was unsuitable for von Neumann's atomic bomb calculations, but he collaborated with the Moore School group to design a successor machine, the EDVAC (Electronic Discrete Variable Automatic Computer). In June 1945, he summarized the group's findings in his *First Draft of a Report on the EDVAC*. The report gave a logical description of what came to be known as the 'stored-program computer', on which almost all subsequent computer developments have been based. The stored-program computer was so called because both the program and the numbers in a calculation shared the same electronic memory, which greatly increased the power and flexibility of the computer: for example, it meant that a program could manipulate its own instructions.

In 1946, von Neumann returned to the Institute for Advanced Study, where he led the construction of one of the first practical computers. With the arrival of working computers, he became interested in areas such as numerical weather forecasting and, more philosophically, in cybernetics and automata. Meanwhile, he continued consulting at Los Alamos, advising on the development of the hydrogen bomb. In 1954 President Eisenhower appointed him to the Atomic Energy Commission, where he had a hawkish influence on science and military policy. In 1955 he was diagnosed with the bone cancer that eventually killed him. His last major activity was preparing the Silliman Lectures for Yale University, which were published posthumously as *The Computer and the Brain* in 1958. He died in 1957 at the age of fifty-three.

Louis Leakey and Mary Leakey

The origins of humankind

(1903–1972 and 1913–1996)

*Louis often used to speak of 'Leakey Luck' when referring
to his and his family's discoveries. But was it luck?
There was the patient ... often backbreaking search;... the
disregard of personal comfort;... the unconventional way
of life;... above all, there was the rugged independence
of outlook, the self-assurance and individualism, the
indefatigable energy and ceaseless drive.*

PHILLIP V. TOBIAS, *A CENTENNIAL TRIBUTE
TO LOUIS LEAKEY*, 2003

Sometimes great scientists make their best discoveries when they
are in a partnership. That was true for Marie and Pierre Curie, who
complemented each other's studies on radioactivity, and it was true
for Louis and Mary Leakey, who pioneered the study of human
origins in East Africa. Working together on numerous expeditions
and excavations, they made the Leakey name synonymous with the
study of human evolution, and proved beyond doubt that humans
emerged first in Africa. Louis and Mary Leakey traced the origin of
the human family tree across an eighteen-million-year span, from
our apelike ancestors to ancient forms of *Homo sapiens*. And they
discovered and named the first human toolmaker, *Homo habilis*. They
made findings independently of each other, too. Mary uncovered a
trail of fossil footprints that showed humans walked upright three
million years ago – about one million years before our ancestors first
began shaping stone tools. And, Louis, the visionary of the pair, helped

launch the first long-term field studies of wild primates, sending Jane Goodall off to record the behaviours of chimpanzees, Dian Fossey to watch mountain gorillas, and Birute Galdikas to observe orang-utans – research that has helped shaped our understanding of the social lives and culture of our early ancestors. Together, Louis and Mary transformed paleoanthropology from a simple hunt for stones and bones into the rich and sophisticated science it is today.

Louis Leakey was the primary instigator of their work, even though Mary made many of their most important discoveries. But it was solely his idea to ignore the prevailing scientific wisdom and search for the bones of our early ancestors in Africa. At the time, paleoanthropologists believed that humans had evolved in Europe and Asia, and only later migrated to Africa. Louis turned that notion on its head, and in time, with Mary's help, completely did away with the old orthodoxy.

Louis's bias in favour of Africa stemmed partly from his roots. He was born to missionary parents who lived among the Kikuyu people in a village in the mountains above Nairobi, British East Africa (now Kenya). Although his parents were British, Louis always considered himself more of a Kikuyu. He was the first white child born among the Kikuyu, and they welcomed him into their lives. When he turned eleven, he joined in the tribe's secret initiation ceremonies for other boys of his age, becoming a member of the Mukanda ('the time of the new robes') group. His parents arranged tutors for him and his two older sisters and younger brother, but overall his schooling was unstructured. Louis had plenty of time to join the far more interesting activities and adventures of his Kikuyu blood brothers. He learned their language, to hunt with bow and arrow, to build traps and track game, and even to catch small prey with his bare hands. Later, he would attribute his seemingly sudden insights into the ways of early humans to his Kikuyu education.

Discovering ancient Africa

And yet it was a children's book he received from a cousin one Christmas that set him on his career path. Entitled *Days Before History*, it recounted the adventures of a young Stone Age British boy named Tig. There were drawings and descriptions of Stone Age men, and the stone tools they made. Inspired, Louis began collecting bits of obsidian he found in the eroded gullies near his home. His family teased him about his 'broken bottles', but Louis had an independent streak. He showed his collection to the only scientist he knew, the curator of a small natural history museum in Nairobi, Arthur Loveridge. Loveridge looked them over and announced that some were 'certainly implements', and explained that little was known about the Stone Age in Africa. With those words, Louis's world changed; he now had a lifelong quest. 'I firmly made up my mind that I would go on until we knew all about the [African] Stone age', he wrote in *White African*, his autobiography. He had just turned thirteen.

Louis's chosen career was not easily attained. He had received only a few years of schooling in England during leaves, but with hard work managed to make up for his academic shortcomings, and was admitted to St John's College, Cambridge. After graduating with a double first in anthropology and modern languages (one of which was Kikuyu), he was awarded a small research fellowship. With it, he booked passage on a ship to Kenya, where in the summer of 1926 he launched his first East African archaeological expedition. One Cambridge professor had tried to dissuade him, telling Louis that he was wasting his time searching for early man in Africa, since 'everyone knew he had started in Asia'. Such naysaying only made Louis more determined to find the evidence that would prove the professor wrong.

Louis ultimately led four East African expeditions. With each one, he pushed further back into the continent's little-known prehistoric eras, uncovering skeletal remains and stone tools that hinted at a past few scientists had ever imagined. He hoped above all that his team of Cambridge students and Kikuyu helpers would find tools like those of the Chellean culture, named after stone hand axes found near Chelles, France. At the time, archaeologists believed the large, teardrop-shaped axes represented the world's oldest culture. In 1929, on Louis's second expedition, John Solomon, the team's geologist, picked up just such an axe near a site called Kariandusi. He was doubtful about his find, but Louis, characteristically, was not. He sent Solomon and a student back to find more, which they did. In those days, there was no way to date the geological strata in which fossils and artefacts were found. Geologists extrapolated the ages of such things by measuring the depth of surrounding sediments, which were thought to accumulate at a steady rate. Using this measure, Louis estimated that the hand axes were about 50,000 years old. Later, using more accurate dating tools, scientists discovered they were closer to 500,000 years old.

Finding tools in Africa that were as old as those in Europe was startling, and Louis was rewarded with enough funds to stage his biggest expedition yet. In 1931, he set out for Olduvai Gorge in Tanganyika Territory (now Tanzania). A twenty-five-mile-long gorge in the Rift Valley, Olduvai snakes its way across the Serengeti Plains, cutting deep into the Earth. Hans Reck, a German geologist, had explored the gorge in 1913, finding an abundance of fossils of extinct mammals as well as the bones of a modern human. Louis read Reck's reports, and although Reck had not found any stone tools in the gorge, Louis had a hunch that the geologist had simply missed them. He invited the German to join his expedition. With four vehicles and a crew of eighteen, they travelled overland from Nairobi, following the rough tracks of Indian traders for three days, until these disappeared. They then bounced along at a painful five miles an hour for another two days, finally reaching the edge of Olduvai Gorge on the morning of 27 September. Shortly after dawn the following day, Louis hiked down into the gorge alone and picked up a hand axe. 'I was nearly mad with delight', he wrote later, 'and rushed back with it into camp', where he awakened the others so they could share in his joy.

Smitten with archaeology

It was after this expedition that Louis met Mary. She was then Mary Nicol, a young artist and aspiring archaeologist. Louis was married, the father of a daughter and an as-yet-unborn child, and virtually penniless. He had some income from the anthropology and archaeology courses he taught at St John's, and he hoped to augment his funds through a popular book about his discoveries, *Adam's Ancestors*. He needed someone to draw the book's illustrations of stone tools, and a friend introduced him to Mary at a dinner party.

The daughter of a landscape painter, Mary had grown up travelling through Italy, Switzerland and France. Like Louis, she had been smitten with archaeology as a child. A French archaeologist guided her and her father into the painted chambers of the prehistoric Pech Merle Cave, then let them search for bits of stone tools in the debris of his excavations. The exploration lit a fire in her. 'After that, I don't think I ever really wanted to do anything else', she said.

Mary also lacked formal schooling. After her father's sudden death, her mother sent her to a convent school, but she managed to get expelled after staging a fit (she put soap in her mouth), and then setting off an explosion in her chemistry class. The explosion 'was quite loud and quite a lot of nuns came running, which will have been good for some of them', she wrote about this incident. Later, she audited archaeology and geology courses at University College London and the Museum of London, and joined various excavations in England as a volunteer.

At twenty, she was unconventional, artistic and witty, a glider pilot with a penchant for French cigarettes. It is not known if she shared all these details with Louis over that first dinner, but the two were drawn to each other, and soon fell deeply in love. Louis invited her on his fourth (and last) East African archaeological expedition, which headed back to Olduvai in January 1935. They took a new route, travelling up a long, muddy track to the summit of Ngorongoro Crater, then down towards the Serengeti and the dark, narrow lines of the gorge. On the plains, there were herds of game, elephant and zebras, rhino and buffalo, and Mary fell in love again, this time with Africa. Together, Louis and Mary scoured the gorge, picking up stone tools and collecting exquisitely preserved fossils of ancient, now-extinct mammals. There were plenty of hand axes, and even more primitive tools that they later called the Oldowan culture (and which are now known to be about two million years old, some of the world's oldest artefacts). But they found only two pieces of bone from an early human skull.

Another twenty years would pass before they found the evidence that would prove Louis right about human origins. During that time, he divorced his first wife; he and Mary wed and had four children, three sons and a daughter who died as an infant. They settled in Nairobi, where Louis became director of the museum in which he had met his first mentor, Loveridge. And they spent every spare moment and penny searching for stones and bones in sites in Kenya and Tanzania.

Sometimes they made spectacular finds. In 1942, at Olorgesailie, a site in the Rift Valley south of Nairobi, they found areas that were literally paved with hand axes, as if early humans had once had a factory for turning them out. And in 1948, on Rusinga Island in Lake Victoria, Mary found the beautifully preserved skull and face-bones of an ancient, twenty-million-year-old ape, *Proconsul* – the first such ape face ever discovered. They found this fossil with the financial assistance of an American businessman based in London, Charles Boise. He continued to provide them with small funds for their expeditions, and in 1959, his aid and their persistence finally paid off at Olduvai. Mary again made the discovery. Louis lay ill in camp, so she went out alone, and began slowly making her way up a rocky incline near the bottom of the gorge. At about eleven o'clock, she noticed a scrap of bone that 'was not lying loose on the surface but projecting from beneath. It seemed to be part of a skull.' She gently brushed away the earth and saw that there were two large teeth set in the curve of a jaw. She jumped into their Land Rover, and drove madly back to camp.

'I've got him! I've got him! I've got him!' she cried. 'Got what?' Louis asked. 'Him, the man! *Our* man. The one we've been looking for. Come quick, I've found his teeth!' Louis quickly regained his health, and together they rushed back to the site. Mary was right: they had finally found their man. Louis at first named the skull *Zinjanthropus*, for man from East Africa. But, later, it was classified as a robust form of *Australopithecus*, an early hominin also known from South Africa. To Louis and Mary, the skull was simply 'Dear Boy'.

A famous family of fossil hunters

In the wake of Dear Boy, Louis and Mary became famous. With new dating technologies, geochronologists were able finally to assign ages to the fossils at Olduvai – and Dear Boy proved to be 'old, old, old', as one of these scientists told Louis. Indeed, the skull was 1.75 million years old – an earth-shaking age that tripled the amount of time that scientists believed humans had existed on the planet. The discovery made headlines around the world, and sparked the beginning of a paleoanthropological bone rush, with scientists hurrying to East Africa to stake their claims. 'The discovery of Zinj [is] the event that opened the present modern era of the truly scientific study of the evolution of man', said Clark Howell, an American paleoanthropologist and colleague of the Leakeys.

The Leakeys were also able to begin full-time excavations at Olduvai with grants from the National Geographic Society. Mary directed these, engaging a crew of Kamba workers, many of whom would become famous fossil hunters in their own right. The digs were often family affairs, with Louis, Mary and their sons Jonathan, Richard and Philip all participating. And it was Jonathan who found the first bits of bone of a new human ancestor, *Homo habilis*, or the 'handy man', the hominin Louis and Mary believed made the earliest, most primitive tools in the gorge.

From the beginning, *Homo habilis* created controversy. If it was a different creature from Dear Boy, it meant that two species of hominins – upright, two-legged early humans – had lived at the same time on the African savannah. Louis argued that such a scenario made perfect sense; one only had to look at the other animals to see that there were numerous varieties of antelopes and primates also living side by side. But many of his fellow scientists strongly criticized this idea of a bushy family tree; they expected to see one long, straight line of ancestors.

And then came vindication. The Leakeys' second son, Richard, had launched his own hominin-hunting expeditions around Kenya's Lake Turkana. Here they found the same pattern: two different species of hominins, one with fairly large bones, the other more gracile, but larger-brained, living side by side. 'They won't believe you!' Louis said when Richard handed him a *Homo habilis* skull from Turkana. But, in time, they did. Today, paleoanthropologists draw various forms of the human family tree – and they are all bushy.

Louis died from a massive heart attack in 1972, one week after seeing Richard's *Homo habilis* skull. He was sixty-nine. Mary continued her excavations at Olduvai. Her team uncovered many more fossils, and thousands of stone tools, all of which she mapped in great detail, producing a nearly two-million-year record of the habitat, animals and humans who once called the gorge home. In 1974, she turned her attention to another site, Laetoli, which had fossils older than those at Olduvai. It was here in 1978 that a team member spotted a trail of unmistakably ancient human footprints – left behind over three million years ago as three individuals walked through the rain while a nearby volcano erupted. As she excavated one of the best of the prints, Mary sat back to admire it. Lighting a cigar, she announced, 'Now this is really something to put on the mantelpiece.'

She was sixty-five when she made that discovery. She continued with her research at Laetoli and Olduvai until the late 1980s. At the time of her death in 1996, Mary was the world's most famous woman archaeologist. She and Louis had accomplished what they set out to achieve: they brought to light the evidence that humans first evolved in Africa. Like all great scientists, they shattered old ideas and ways of thinking – and they did it with nothing more than stones and bones.

POSTSCRIPT
Science and scientists in our time

Albert Einstein is probably the most quoted (and misquoted) scientist in history. The website Wikiquote has many more entries for him than for Aristotle, Galileo Galilei, Isaac Newton, Charles Darwin, Marie Curie or Stephen Hawking, and even for Einstein's opinionated contemporaries Winston Churchill and George Bernard Shaw.

I therefore make no apology for revisiting an Einstein comment (quoted in the introduction to 'Inside the Atom') that is highly relevant to 21st-century science, at the close of this book about five centuries of key scientists: 'Science is not and never will be a closed book. Every important advance brings new questions. Every development reveals, in the long run, new and deeper difficulties.' This dates from 1938, in Einstein's first book for a lay readership, *The Evolution of Physics,* a collaboration between the German-speaking genius and Leopold Infeld, a fellow physicist who was fluent in English.

In physics, Einstein's theory of relativity and the quantum theory developed by Einstein, Niels Bohr, Werner Heisenberg, Erwin Schrödinger and many subsequent physicists certainly raised 'new questions' and required more specialized knowledge to comprehend than, say, Newton's theory of motion and gravity and Michael Faraday's theory of electricity and magnetism. Even Einstein experienced 'new and deeper difficulties' with one of general relativity's key astronomical predictions. The black hole was an idea first proposed as a 'dark star' by the natural philosopher (and clergyman) John Michell as early as 1783, based on Newtonian mechanics. Yet for Einstein, this theoretical concept – a sharply curved four-dimensional structure precisely delineated by his equations – was too bizarre to accept. He controversially repudiated the black hole's existence in a famous paper of 1939, published by the leading US mathematical journal *Annals of Mathematics*, and continued to do so until his death in 1955, for reasons he never fully explained. Not until 2017 was a black hole directly observed, by the Event Horizon Telescope, and a sensational image of it published two years later. In 2020, mathematical physicist Roger Penrose shared the Nobel prize for 'the discovery that black hole formation is a robust prediction of the general theory of relativity'.

Since Einstein's time, in response to the increasing complexity of science he predicted, most researchers have felt obliged to become more specialized. Far fewer now embrace several disciplines like the polymath Thomas Young, the naturalist, geologist and biologist Charles Darwin and the climatologist, geologist, geophysicist, meteorologist and polar researcher Alfred Wegener. Instead, modern researchers choose to collaborate with other specialists more extensively, by forming large groups often known as 'research clusters'. Scientific advances in the 21st century are therefore less the product of solitary individuals than in the past, when many scientists worked almost alone – epitomized by the self-isolating Newton in his Cambridge college rooms or the youthful Einstein in Switzerland. Today, it is common for a scientific paper in a journal such as *Nature*, *Science* or *The Lancet* to be credited to a dozen (and frequently many more) authors, each contributing his or her limited expertise, rather than being authored by one or two scientists, as was generally true of Einstein's published papers.

Consider the following celebrated examples of current science, too recent to have been covered by this book, which is deliberately restricted to deceased scientists (apart from Francis Crick's collaborator James Watson):

1. The exploration of space, inaugurated by the *Sputnik* satellite in 1957 and typified by the Hubble Space Telescope launched in 1990, has been organized by national and international agencies, most notably the space agency NASA, not by individual scientists.

2. Global research in climate science has been developed by the United Nations' Intergovernmental Panel on Climate Change founded in 1988 – unlike the first accurate monitoring of increasing atmospheric carbon dioxide from 1958, the origin of the Keeling Curve, which was devised by a single scientist, the geochemist Charles David Keeling.

3. The development of the internet in the 1990s was the work of a large group: the World Wide Web Consortium, today counting over 450 member organizations – unlike the development of the digital computer by Alan Turing and, separately, John von Neumann from the 1930s to the 1950s.

4. The draft sequence of the human genome announced in 2000 was the work of another group: the Human Genome Project, involving 20 major institutions, companies and laboratories – in contrast with the discovery of DNA's structure by Rosalind Franklin, Crick and Watson in the early 1950s.

5. The elementary particle known as the Higgs boson that supports the current Standard Model of particle physics was discovered by the multinational Large Hadron Collider project at the CERN laboratory in 2012 – not by the theoretical physicist Peter Higgs, who predicted its existence in 1964.

6. During the worldwide pandemic beginning in 2020, the names of pharmaceutical companies have dominated the news, rather than the names of the many epidemiologists researching and testing COVID-19 vaccines – unlike earlier breakthroughs such as the rabies vaccine devised by the chemist and microbiologist Louis Pasteur in 1855 and the polio vaccine developed by the virologist Jonas Salk in 1955.

That said, outstanding individuals unquestionably remain pivotal to scientific progress, as indicated by the annual award of Nobel prizes in chemistry, physics and physiology or medicine, which are still restricted to a maximum of three scientists per award and never given to large research groups. The computer scientist Tim Berners-Lee, who founded the World Wide Web Consortium in 1994, was key to the development of the internet, as was the biotechnologist Craig Venter to the sequencing of the human genome, and the Keeling Curve of rising carbon dioxide concentration to the understanding of climate change. But whether such scientists of our time belong in the same league as Turing, Crick and Wegener is still uncertain. It is too soon to be sure of their long-term status – not least because some, such as Berners-Lee (who received the Turing Award for inventing the World Wide Web in 2016), are still active researchers. However, of one thing we may probably be certain: all of the greatest scientists, past and present, would agree with Newton's verdict on how he derived his most famous scientific achievement: 'By thinking on it continually.'

CONTRIBUTORS

ANDREW ROBINSON is the author of more than twenty-five books on science, the history of science and the arts, such as the award-winning *Earthshock: Hurricanes, Volcanoes, Earthquakes, Tornadoes and Other Forces of Nature*; *The Story of Measurement*; and *Genius: A Very Short Introduction*. He has written biographies of Albert Einstein (*A Hundred Years of Relativity* and *Einstein on the Run*), Thomas Young (*The Last Man Who Knew Everything*), Michael Ventris (*The Man Who Deciphered Linear B*) and Jean-François Champollion (*Cracking the Egyptian Code*). A King's Scholar of Eton College, he holds degrees from Oxford University (in chemistry) and the School of Oriental and African Studies, London, and was a visiting fellow of Wolfson College, Cambridge from 2006 to 2010. Having been literary editor of the *Times Higher Education Supplement* for twelve years, in 2007 he became a full-time writer and journalist. He reviews regularly for newspapers and magazines, including the science journals *Nature*, *Science* and *The Lancet*. **Albert Einstein, Marie Curie and Pierre Curie**

JIM AL-KHALILI is professor of theoretical physics and chair in the public engagement in science at the University of Surrey. He has presented BBC productions about science and written several popular books, including *Quantum: A Guide for the Perplexed*; *Nucleus: A Trip into the Heart of Matter*; and *Pathfinders: The Golden Age of Arabic Science*. **Hideki Yukawa**

JONATHAN BOWEN is emeritus professor of computing at London South Bank University. His research interests cover the history of computing, museum informatics and software engineering. As well as numerous technical papers, his publications include contributions to the *Encyclopedia of Computers and Computer History* and *The Oxford Companion to the History of Modern Science*. **Alan Turing**

NATHAN BROOKS is associate professor of history at New Mexico State University. A former Fulbright Scholar and fellow of the Chemical Heritage Foundation, he has researched and published on the Russian history of chemistry, focusing on the life and work of Dmitri Mendeleev. **Dmitri Mendeleev**

HELEN BYNUM was formerly a lecturer in medical history at the University of Liverpool and is now an independent editor, writer and lecturer. She is the author of *Spitting Blood: A History of Tuberculosis* and the editor of the multi-volume *Dictionary of Medical Biography* and *Great Discoveries in Medicine* (both with William F. Bynum). **Andreas Vesalius**

MARTIN CAMPBELL-KELLY is emeritus professor of computer

science at the University of Warwick. He is on the editorial board of the IEEE Annals of the History of Computing. His books include *Computer: A History of the Information Machine* (with William Aspray) and *From Airline Reservations to Sonic the Hedgehog: A History of the Software Industry*. **John von Neumann**

JORDI CAT is associate professor of the history and philosophy of science at Indiana University, Bloomington. His books include *Maxwell, Sutton, and the Birth of Color Photography* (2013) and the forthcoming *Master and Designer of Fields: James Clerk Maxwell and Constructive, Connective, Conventionalist, and Concrete Natural Philosophy*. **James Clerk Maxwell**

FRANK CLOSE is emeritus professor of theoretical physics at Oxford University, and a fellow of Exeter College, Oxford. His research focuses on the quark and gluon structure of strongly interacting particles. He is the author of several popular books, including *The Particle Odyssey* (with Michael Marten and Christine Sutton); *Antimatter*; *Neutrino*; and *The Infinity Puzzle*. **Ernest Rutherford**

GEORGINA FERRY is a science writer, author and broadcaster based in Oxford. Her books include *Max Perutz and the Secret of Life* and *Dorothy Hodgkin: A Life*, which was shortlisted for the Duff Cooper Prize and the Marsh Biography Award. **Dorothy Crowfoot Hodgkin**

LEON FINE is the vice-dean for research and chair of the department of biomedical sciences at the Cedars-Sinai Medical Center in Los Angeles. He has written and edited several publications on the history and practice of medicine, including *The Young Harvey* and *Harvey's Keepers: Harveian Librarians through the Ages*. **William Harvey**

RONALD FISHMAN was a retired ophthalmologist and historian of ophthalmology, based in Washington, DC. As well as regularly lecturing and publishing papers on the subject, he also wrote on the life and work of Santiago Ramón y Cajal and Camillo Golgi. **Santiago Ramón y Cajal**

TORE FRÄNGSMYR was the Hans Rausing Professor in the History of Science at Uppsala University. He was director of the Centre for History of Science at the Royal Swedish Academy of Sciences, and was editor between 1988 and 2008 of Les Prix Nobel, the yearbook of the Nobel Foundation. He published some twenty-five books, including *Linnaeus: The Man and His Work* and *Alfred Nobel*. **Carl Linnaeus**

NICHOLAS GILLHAM was the James B. Duke Professor of Biology at Duke University. An expert in the field of genetics and its history, he wrote *Organelle Genes and Genomes*; *Genes, Chromosomes, and Disease*; and *A Life of Sir Francis Galton: From African Exploration to the Birth of Eugenics*. **Francis Galton**

MICHAEL HUNTER is emeritus professor of history at Birkbeck College, University of London. He is the editor of the collected works of Robert Boyle (with Edward B. Davis) and of Boyle's correspondence (with Antonio Clericuzio and Lawrence M. Principe). His other books include *Boyle: Between God and Science* and *The Boyle Papers: Understanding the Manuscripts of Robert Boyle*.
Robert Boyle

ROB ILIFFE is professor of history of science at the University of Oxford. He is editorial director of the online Newton Project and director of the AHRC Newton Theological Papers Project, and the author of *Newton: A Very Short Introduction*. He is also co-editor of *Annals of Science* and was formerly editor of the journal *History of Science*. **Isaac Newton**

FRANK A. J. L. JAMES is professor of the history of science at University College London and was formerly head of collections at the Royal Institution, London. He is the author of *Michael Faraday: A Very Short Introduction* and published a complete edition of Faraday's correspondence from 1991 to 2011.
Michael Faraday

GEERDT MAGIELS is an independent biologist and philosopher of science. He has published on science, medicine, psychiatry and art, and is the author of *From Sunlight to Insight: Jan IngenHousz, the Discovery of*

Photosynthesis and Science in the Light of Ecology. **Jan IngenHousz**

ROGER McCOY is emeritus professor of physical geography at the University of Utah and the author of *Ending in Ice: The Revolutionary Idea and Tragic Expedition of Alfred Wegener* and *On the Edge: Mapping North America's Coasts*.
Alfred Wegener

PATRICK MOORE was an astronomer, and a writer, researcher, radio commentator and television presenter. A fellow of the Royal Society and a former president of the British Astronomical Association, he wrote more than 100 books on astronomy. From 1957 he presented the world's longest-running television series with the same original presenter, *The Sky at Night*, on the BBC.
Edwin Powell Hubble

VIRGINIA MORELL is a science journalist, author and lecturer. A correspondent for *Science* and a regular contributor to *National Geographic*, she specializes in the fields of evolutionary and conservation biology. Her biography of the Leakeys, *Ancestral Passions: The Leakey Family and the Quest for Humankind's Beginnings*, was a Notable Book of the Year in the *New York Times*.
Louis Leakey and Mary Leakey

ROBERT OLBY is research professor in the department of history and philosophy of science at the University of Pittsburgh. He has published and lectured widely on biology, genetics and molecular biology, and is the author of *Origins of Mendelism*; *Francis Crick: Hunter of Life's Secrets*; and *The Path to the Double Helix*. **Francis Crick and James Watson**

ROBERT PARADOWSKI is an emeritus professor in the department of science, technology and society/public policy at the Rochester Institute of Technology. His research on the life and work of Linus Pauling began with his dissertation, *The Structural Chemistry of Linus Pauling*, in 1972, and has continued with many articles in journals, books and reference works. **Linus Pauling**

JAY PASACHOFF is director of the Hopkins Observatory and Field Memorial Professor of Astronomy at Williams College in Williamstown, Massachusetts. He is the author of books on astronomy, physics, mathematics and other sciences, including *The Solar Corona* (with Leon Golub); *The Cosmos: Astronomy in the New Millennium* (with Alex Filippenko); and *Fire in the Sky: Comets and Meteors, the Decisive Centuries, in British Art and Science* (with Roberta J. M. Olson). **Nicolaus Copernicus, Johannes Kepler, Galileo Galilei**

NAOMI PASACHOFF is a research associate at Williams College in Williamstown, Massachusetts. She has written more than twenty science textbooks and biographies of scientists, including *Marie Curie and*

the Science of Radioactivity; *Niels Bohr: Physicist and Humanitarian*; and *Linus Pauling: Advancing Science, Advocating Peace*. **Nicolaus Copernicus, Johannes Kepler, Galileo Galilei**

ALISON PEARN is associate director of the Darwin Correspondence Project at the University of Cambridge. She regularly gives lectures on Darwin, writes for the popular press, and takes part in broadcasts on the subject. She has jointly edited nine volumes of Darwin's letters and is the author of *A Voyage Round the World: Charles Darwin and the Beagle Collections of the University of Cambridge*. **Charles Darwin**

JEAN-PIERRE POIRIER has both a medical degree and a doctorate in economics. Formerly director of research at a French pharmaceutical company, he is a member of the Comité Lavoisier at the French Academy of Sciences and author of *Lavoisier: Chemist, Biologist, Economist*. **Antoine-Laurent Lavoisier**

ALAN ROCKE is Henry Eldridge Bourne Professor of History Emeritus at Case Western Reserve University in Cleveland, Ohio. Specializing in the history of the physical sciences during the 19th and 20th centuries, he is the author of many publications on the history of science in Germany and France, including *Image and Reality: Kekulé, Kopp, and the Scientific Imagination*. **John Dalton, August Kekulé**

MARTIN RUDWICK is emeritus professor in history at the University of California, San Diego and a research scholar in the department of history and philosophy of science at the University of Cambridge. His main field of study is the history of the earth sciences, for which he has received the Sue Tyler Friedman Medal and the George Sarton Medal. His books include *The Meaning of Fossils*; *The Great Devonian Controversy*; *Scenes from Deep Time*; and most recently *Bursting the Limits of Time* and its sequel *Worlds Before Adam*. **James Hutton, Charles Lyell**

GINO SEGRE is emeritus professor of physics and astronomy at the University of Pennsylvania. He is author of several academic and popular books about science, including *Faust in Copenhagen: A Struggle for the Soul of Physics and the Birth of the Nuclear Age*; *A Matter of Degrees* (published in the UK as *Einstein's Refrigerator*); and *Ordinary Geniuses*. **Enrico Fermi**

VIRENDRA SINGH was director of the Tata Institute for Fundamental Research in Mumbai, from 1987 to 1997. He was subsequently C. V. Raman Distinguished Professor of the Indian National Science Academy. He has published numerous technical papers on theoretical physics and quantum mechanics. **Chandrasekhar Venkata Raman**

MARK SOLMS is a psychoanalyst and honorary lecturer in neurosurgery at the St Bartholomew's and Royal London School of Medicine, chair of neuropsychology at the University of Cape Town, South Africa, and director of the Arnold Pfeffer Center for Neuropsychoanalysis at the New York Psychoanalytic Institute. He has won numerous awards for his work in neuropsychoanalysis. His books include *The Brain and the Inner World*. **Sigmund Freud**

NICHOLAS WADE is emeritus professor in psychology at the University of Dundee. His writings on vision, perception and their relationship with fields such as art include *Purkinje's Vision: The Dawning of Neuroscience* (with Josef Brožek); *A Natural History of Vision*; *Visual Perception: An Introduction* (with Michael Swanston); *Perception and Illusion*; and *Circles: Science, Sense and Symbol*. **Jan Purkinje**

LAURA DASSOW WALLS is the William P. and Hazel B. White Professor of English at the University of Notre Dame, Indiana. Her books include the award-winning *The Passage to Cosmos: Alexander von Humboldt and the Shaping of America and Seeing New Worlds: Henry David Thoreau and Nineteenth-Century Natural Science*. **Alexander von Humboldt**

ANDREW WHITAKER is emeritus professor of physics at Queen's University, Belfast. He is the author of several books on the foundations of quantum theory, including *Einstein, Bohr and the Quantum Dilemma*; *The New Quantum Age: From Bell's Theorem to Quantum Computation and Teleportation*; and *Einstein's Struggles with Quantum Theory: A Reappraisal* (with Dipankar Home). **Niels Bohr**

ROGER WOOD is honorary reader in genetics in the faculty of life sciences at the University of Manchester. He is the author of a number of publications on the life and work of Gregor Mendel, including *Genetic Prehistory in Selective Breeding: A Prelude to Mendel* (with Vítězslav Orel). **Gregor Mendel**

MICHAEL WORBOYS is emeritus professor in the Centre for the History of Science, Technology and Medicine at the University of Manchester. His publications include *Mad Dogs and Englishmen: Rabies in Britain, 1830–2000* (with Neil Pemberton) and *Spreading Germs: Disease Theories and Medical Practice in Britain, 1865–1900*. **Louis Pasteur**

FURTHER READING

UNIVERSE
Nicolaus Copernicus
Owen Gingerich, *The Eye of Heaven: Ptolemy, Copernicus, Kepler* (New York: American Institute of Physics, 1993)
———— and James MacLachlan, *Nicolaus Copernicus: Making the Earth a Planet* (Oxford and New York: Oxford University Press, 2005)
Thomas Kuhn, *The Copernican Revolution* (Cambridge, MA: Harvard University Press, 1957)
Dava Sobel, *A More Perfect Heaven: How Copernicus Revolutionized the Cosmos* (New York: Walker, 2011)
Christopher Walker (ed.), *Astronomy Before the Telescope* (London: British Museum Press, 1996)

Johannes Kepler
Max Caspar, *Kepler*, trans. C. Doris Hellman (New York: Dover, 1993)
Kitty Ferguson, *Tycho and Kepler: The Unlikely Partnership That Forever Changed Our Understanding of the Heavens* (New York: Walker, 2002)
Judith V. Field, *Kepler's Geometrical Cosmology* (Chicago: University of Chicago Press, 1988)
Rhonda Martens, *Kepler's Philosophy and the New Astronomy* (Princeton: Princeton University Press, 2000)
Bruce Stephenson, *The Music of the Heavens: Kepler's Harmonic Astronomy* (Princeton: Princeton University Press, 1994)
James R. Voelkel, *Johannes Kepler and the New Astronomy* (Oxford and New York: Oxford University Press, 1999)

Galileo Galilei
Richard J. Blackwell, *Galileo, Bellarmine, and the Bible* (South Bend, IN: University of Notre Dame Press, 1991)
Stillman Drake, *Galileo at Work* (Chicago: University of Chicago Press, 1978; reprinted New York: Dover, 1995)
————, *Galileo: Pioneer Scientist* (Toronto: University of Toronto Press, 1989)
John Heilbron, *Galileo* (Oxford and New York: Oxford University Press, 2010)
James MacLachlan, *Galileo Galilei: First Physicist* (Oxford and New York: Oxford University Press, 1997)
Michael Sharratt, *Galileo: Decisive Innovator* (Oxford: Blackwell, 1994)

Isaac Newton
Betty Jo Teeter Dobbs, *The Janus Faces of Genius: The Role of Alchemy in Newton's Thought* (Cambridge: Cambridge University Press, 1991)
A. Rupert Hall, *Philosophers at War: The Quarrel between Newton and Leibniz* (Cambridge: Cambridge University Press, 1980)
Rob Iliffe, *Isaac Newton: A Very Short Introduction* (Oxford and New York: Oxford University Press, 2008)
Frank Manuel, *A Portrait of Isaac Newton* (Cambridge, MA: Harvard University Press, 1968)
Richard S. Westfall, *Never at Rest: A Biography of Isaac Newton* (Cambridge: Cambridge University Press, 1984)

Michael Faraday

Geoffrey Cantor, *Michael Faraday: Sandemanian and Scientist. A Study of Science and Religion in the Nineteenth Century* (London: Macmillan, 1991)

David Gooding, *Experiment and the Making of Meaning: Human Agency in Scientific Observation and Experiment* (Dordrecht: Kluwer, 1990)

Bruce J. Hunt, 'Michael Faraday, cable telegraphy and the rise of field theory', *History of Technology*, vol. 13 (1991), pp. 1–19

Frank A. J. L. James, *Michael Faraday: A Very Short Introduction* (Oxford and New York: Oxford University Press, 2010)

————— (ed.), *The Correspondence of Michael Faraday*, 6 vols (London: Institution of Electrical Engineers, 1991–2011)

James Clerk Maxwell

Jordi Cat, *Master and Designer of Fields: James Clerk Maxwell and Constructive, Connective and Concrete Natural Philosophy* (Oxford: Oxford University Press, forthcoming)

C. W. F. Everitt, *James Clerk Maxwell, Physicist and Natural Philosopher* (New York: Scribner's, 1975)

Martin Goldman, *The Demon in the Aether: The Life of James Clerk Maxwell* (Edinburgh: Paul Harris, 1983)

Peter M. Harman, *The Natural Philosophy of James Clerk Maxwell* (Cambridge: Cambridge University Press, 1995)

Crosbie Smith, *The Science of Energy* (Chicago: University of Chicago Press, 1988)

Albert Einstein

Albert Einstein, *Relativity: The Special and the General Theory* (London: Routledge Classics, 2001; first published 1916)

Albrecht Fölsing, *Albert Einstein: A Biography* (London: Viking, 1997)

John S. Rigden, *Einstein 1905: The Standard of Greatness* (Cambridge, MA: Harvard University Press, 2005)

Andrew Robinson, *Einstein: A Hundred Years of Relativity*, rev. edn (Princeton: Princeton University Press, 2015)

Paul Arthur Schilpp (ed.), *Albert Einstein: Philosopher-Scientist*, includes Einstein's 'Autobiographical Notes' (Evanston, IL: Library of Living Philosophers, 1949)

Edwin Powell Hubble

Gale E. Christianson, *Edwin Hubble: Mariner of the Nebulae* (New York: Farrar, Straus & Giroux, 1995)

Mary V. Fox, *Edwin Hubble: American Astronomer* (New York: Franklin Watts, 1997)

Edwin Powell Hubble, *The Realm of the Nebulae* (Oxford: Oxford University Press, 1936)

Patrick Moore, *The Data Book of Astronomy* (Bristol: Institute of Physics Publishing, 2000)

Alexander S. Sharov and Igor D. Novikov, *Edwin Hubble: The Discoverer of the Big Bang Universe*, trans. Vitaly I. Kisin (Cambridge: Cambridge University Press, 1993)

EARTH
James Hutton

Dennis R. Dean, *James Hutton and the History of Geology* (Ithaca: Cornell University Press, 1992)

Stephen Jay Gould, *Time's Arrow, Time's Cycle: Myth and Metaphor in the Discovery of Geological Time* (Cambridge, MA: Harvard University Press, 1987)

Rachel Laudan, *From Mineralogy to Geology: The Foundations of a Science, 1650–1830* (Chicago: University of Chicago Press, 1987)

R. S. Porter, *The Making of Geology: Earth Sciences in Britain 1660–1815* (Cambridge: Cambridge University Press, 1987)

Martin J. S. Rudwick, *Bursting the Limits of Time: The Reconstruction of Geohistory in the Age of Revolution* (Chicago: University of Chicago Press, 2005)

Charles Lyell

D. J. Blundell and A. C. Scott (eds), *Lyell: The Past is the Key to the Present* (London: Geological Society, 1998)

Stephen Jay Gould, *Time's Arrow, Time's Cycle: Myth and Metaphor in the Discovery of Geological Time* (Cambridge, MA: Harvard University Press, 1987)

Rachel Laudan, *From Mineralogy to Geology: The Foundations of a Science, 1650–1830* (Chicago: University of Chicago Press, 1987)

Martin J. S. Rudwick, *Worlds Before Adam: The Reconstruction of Geohistory in the Age of Reform* (Chicago: University of Chicago Press, 2008)

L. G. Wilson, *Charles Lyell: The Years to 1841* (New Haven and London: Yale University Press, 1972)

Alexander von Humboldt

Gerard Helferich, *Humboldt's Cosmos: Alexander von Humboldt and the Latin American Journey that Changed the Way We See the World* (New York: Gotham Books, 2004)

Alexander von Humboldt, *Personal Narrative of a Journey to the Equinoctial Regions of the New Continent*, abr. and trans. Jason Wilson (New York and London: Penguin, 1995)

——— and Aimé Bonpland, *Essay on the Geography of Plants*, ed. Stephen T. Jackson, trans. Sylvie Romanowski (Chicago: University of Chicago Press, 2010)

Aaron Sachs, *The Humboldt Current: 19th-Century Exploration and the Roots of American Environmentalism* (New York: Viking, 2006)

Laura Dassow Walls, *The Passage to Cosmos: Alexander von Humboldt and the Shaping of America* (Chicago: University of Chicago Press, 2009)

Andrea Wulf, *The Invention of Nature: The Adventures of Alexander von Humboldt, the Lost Hero of Science* (London: John Murray, 2015)

Alfred Wegener

Johannes Georgi, *Mid Ice: The Story of the Wegener Expedition to Greenland* (New York: E. P. Dutton, 1935)

Roger M. McCoy, *Ending in Ice: The Revolutionary Idea and Tragic Expedition of Alfred Wegener*

(New York and Oxford: Oxford University Press, 2006)

Naomi Oreskes, *Plate Tectonics: An Insider's History of the Modern Theory of the Earth* (Boulder, CO: Westview, 2001)

Martin Schwarzbach, *Alfred Wegener: The Father of Continental Drift* (Madison, WI: Science Tech Publishers, 1986)

Alfred Wegener, *Origin of Continents and Oceans*, English trans. of 4th edn (New York: Dover, 1966; first published in 1929)

MOLECULES AND MATTER
Robert Boyle

Michael Hunter, *Boyle: Between God and Science* (New Haven and London: Yale University Press, 2009)

————— (ed.), *Robert Boyle Reconsidered* (Cambridge: Cambridge University Press, 1994)

————— and Edward B. Davis (eds), *The Works of Robert Boyle*, 14 vols (London: Pickering and Chatto, 1999–2000)

Lawrence M. Principe, *The Aspiring Adept: Robert Boyle and his Alchemical Quest* (Princeton: Princeton University Press, 1998)

Antoine-Laurent de Lavoisier

Bernadette Bensaude-Vincent, *Lavoisier: Mémoires d'une revolution* (Paris: Flammarion, 1993)

Arthur Donovan, *Antoine Lavoisier: Science, Administration and Revolution* (Cambridge, MA: Blackwell, 1993)

Henry Guerlac, *Lavoisier, The Crucial Year: The Background and Origin of His First Experiments on Combustion in 1772* (Ithaca: Cornell University Press, 1961)

Frederic Lawrence Holmes, *Antoine Lavoisier: The Next Crucial Year or The Sources of His Quantitative Method in Chemistry* (Princeton: Princeton University Press, 1998)

Jean-Pierre Poirier, *Lavoisier: Chemist, Biologist, Economist* (Philadelphia: Pennsylvania University Press, 1996)

Lisa Yount, *Antoine Lavoisier: Founder of Modern Chemistry* (Springfield, NJ: Enslow Publishers, 1997)

John Dalton

Donald Cardwell (ed.), *John Dalton and the Progress of Science* (Manchester: Manchester University Press, 1968)

Frank Greenaway, *John Dalton and the Atom* (Ithaca: Cornell University Press, 1966)

Elizabeth C. Patterson, *John Dalton and the Atomic Theory* (New York: Doubleday, 1970)

Robert Angus Smith, *Memoir of John Dalton and History of the Atomic Theory* (London: Bailliere, 1856)

Arnold Thackray, *John Dalton: Critical Assessments of His Life and Science* (Cambridge, MA: Harvard University Press, 1972)

Dmitri Mendeleev

Igor S. Dmitriev, 'Scientific discovery in *statu nascendi*: the case of Dmitrii Mendeleev's Periodic Law', *Historical Studies in the Physical Sciences*, vol. 34 (2004), pp. 235–75

William B. Jensen (ed.), *Mendeleev on the Periodic Law: Selected Writings, 1869–1905* (Mineola, NY: Dover, 2005)

Michael D. Gordin, *A Well-Ordered Thing: Mendeleev and the Shadow of the Periodic Table* (New York: Basic Books, 2004)

Eric R. Scerri, *The Periodic Table: Its Story and Its Significance* (New York: Oxford University Press, 2007)

J. W. van Spronsen, *The Periodic System of the Chemical Elements: The First One Hundred Years* (Amsterdam: Elsevier, 1969)

August Kekulé

Richard Anschütz, *August Kekulé: Leben und Wirken* (Berlin: Verlag Chemie, 1929)

O. Theodor Benfey (ed.), *Kekulé Centennial* (Washington, DC: American Chemical Society, 1966)

John Buckingham, *Chasing the Molecule* (Stroud: Sutton, 2004)

Alan J. Rocke, *Image and Reality: Kekulé, Kopp, and the Scientific Imagination* (Chicago: University of Chicago Press, 2010)

Colin A. Russell, *The History of Valency* (Leicester: Leicester University Press, 1971)

Dorothy Crowfoot Hodgkin

Guy Dodson, Jenny P. Glusker and David Sayre (eds), *Structural Studies on Molecules of Biological Interest: A Volume in Honour of Dorothy Hodgkin* (Oxford: Clarendon Press, 1981)

Georgina Ferry, *Dorothy Hodgkin: A Life* (London: Granta, 1998; Cold Spring Harbor, NY: Cold Spring Harbor Laboratory Press, 2000)

Dorothy Crowfoot Hodgkin, 'The X-ray analysis of complicated molecules', in *Nobel Lectures, Chemistry 1963–1970* (Amsterdam: Elsevier, 1972)

Sharon B. McGrayne, *Nobel Prize Women in Science: Their Lives, Struggles and Momentous Discoveries*, 2nd edn (Washington, DC: Joseph Henry Press, 2001)

Chandrasekhar Venkata Raman

A. Jayaraman, *C. V. Raman: A Memoir* (New Delhi: Affiliated East-West, 1989)

S. Ramaseshan (ed.), *The Scientific Papers of Sir C. V. Raman*, 6 vols (Bangalore: Indian Academy of Sciences, 1988)

——— and C. Ramachandra Rao (eds), *C. V. Raman: A Pictorial Biography* (Bangalore: Indian Academy of Sciences, 1988)

G. Venkataraman, *Journey into Light: Life and Science of C. V. Raman* (Bangalore: Indian Academy of Sciences, 1988)

INSIDE THE ATOM
Marie Curie and Pierre Curie

Eve Curie, *Madame Curie: A Biography* (New York: Da Capo Press, 2001; first published 1937)

Marie Curie, *Pierre Curie*, with autobiographical notes by Marie Curie (New York: Macmillan, 1923)

Barbara Goldsmith, *Obsessive Genius: The Inner World of Marie Curie* (London: Phoenix, 2005)

Susan Quinn, *Marie Curie: A Life* (London: Heinemann, 1995)

Alfred Romer (ed.), *The Discovery of Radioactivity and Transmutation* (New York: Dover, 1964)

Ernest Rutherford

John Campbell, *Rutherford: Scientist Supreme* (Christchurch, NZ: AAS Publications, 1999)

Brian Cathcart, *The Fly in the Cathedral: How a Small Group of Cambridge Scientists Won the Race to Split the Atom* (London: Viking, 2004)

Frank Close, *Particle Physics: A Very Short Introduction* (Oxford and New York: Oxford University Press, 2004)

————, Michael Marten and Christine Sutton, *The Particle Odyssey: A Journey to the Heart of Matter* (Oxford and New York: Oxford University Press, 2002)

David Wilson, *Rutherford: Simple Genius* (London: Hodder & Stoughton, 1983)

Niels Bohr

Finn Aaserud, *Redirecting Science: Niels Bohr, Philanthropy, and the Rise of Nuclear Physics* (Cambridge: Cambridge University Press, 1990)

H. J. Folse, *The Philosophy of Niels Bohr* (Amsterdam: North-Holland, 1985)

Michael Frayn, *Copenhagen* (London: Methuen, 1998)

Ruth E. Moore, *Niels Bohr: The Man, His Science and the World He Changed* (New York: Alfred A. Knopf, 1966)

Abraham Pais, *Niels Bohr's Times in Physics, Philosophy and Polity* (Oxford and New York: Oxford University Press, 1991)

Andrew Whitaker, *Einstein, Bohr and the Quantum Dilemma*, 2nd edn (Cambridge: Cambridge University Press, 2006)

Linus Carl Pauling

Ted Goertzel and Ben Goertzel, *Linus Pauling: A Life in Science and Politics* (New York: Basic Books, 1995)

Thomas Hager, *Force of Nature: The Life of Linus Pauling* (New York: Simon & Schuster, 1995)

Barbara Marinacci (ed.), *Linus Pauling in His Own Words: Selections from His Writings, Speeches, and Interviews* (New York: Simon & Schuster, 1995)

Fumikazu Miyazaki (ed.), *Linus Pauling: A Man of Intellect and Action* (Tokyo: Cosmos Japan International, 1991)

John W. Servos, *Physical Chemistry from Ostwald to Pauling: The Making of a Science in America* (Princeton: Princeton University Press, 1990)

Enrico Fermi

James Cronin (ed.), *Fermi Remembered* (Chicago: University of Chicago Press, 2004)

Laura Fermi, *Atoms in the Family* (Chicago: University of Chicago Press, 1954)

George Gamow, *Thirty Years That Shook Physics* (New York: Dover, 1985)

Abraham Pais, *Inward Bound* (Oxford and New York: Oxford University Press, 1986)

Richard Rhodes, *The Making of the Atomic Bomb* (New York: Simon & Schuster, 1988)

Emilio Segrè, *Enrico Fermi, Physicist* (Chicago: University of Chicago Press, 1970)

Hideki Yukawa

Ioan James, *Remarkable Physicists: From Galileo to Yukawa* (Cambridge: Cambridge University Press, 2004)

N. Kemmer, 'Hideki Yukawa', *Biographical Memoirs of Fellows of the Royal Society*, vol. 29, pp. 661–76 (London: Royal Society, 1983)

Humitaka Sato, 'Biography of Hideki Yukawa', 'Proceedings of the 23rd International Nuclear Physics Conference', *Nuclear Physics A*, vol. 805 (2008), pp. 21c–28c

Hideki Yukawa, *Creativity and Intuition: A Physicist Looks at East and West*, trans. John Bester (New York: Kodansha International, 1973)

LIFE

Carl Linnaeus

Wilfrid Blunt, *The Complete Naturalist: A Life of Linnaeus* (New York: Viking, 1971)

Tore Frängsmyr (ed.), *Linnaeus: The Man and His Work* (Canton, MA: Science History Publications, 1994)

Lisbet Koerner, *Linnaeus: Nature and Nation* (Cambridge, MA: Harvard University Press, 1999)

James L. Larson, *Interpreting Nature: The Science of Living Forms from Linnaeus to Kant* (Baltimore: Johns Hopkins University Press, 1994)

————, *Reason and Experience: The Representation of Natural Order in the Work of Carl von Linné* (Berkeley: University of California Press, 1971)

John Weinstock (ed.), *Contemporary Perspectives of Linnaeus* (Lanham, MD: University Press of America, 1985)

Jan IngenHousz

Norman Beale and Elaine Beale, *Echoes of Ingen Housz: The Long Lost Story of the Genius Who Rescued the Habsburgs from Smallpox and Became the Father of Photosynthesis* (Salisbury: Hobnob Press, 2011)

————, 'Evidence-based medicine in the eighteenth century: the Ingen Housz-Jenner correspondence revisited', *Medical History*, vol. 1 (2005), pp. 49, 79–98

Howard Gest, 'Bicentenary homage to Dr Jan Ingen-Housz, MD (1730–1799), pioneer of photosynthesis research', *Photosynthesis Research*, vol. 63 (2000), pp. 183–190

Geerdt Magiels, *From Sunlight to Insight: Jan IngenHousz, the Discovery of Photosynthesis and Science in the Light of Ecology* (Brussels: VUB Press, 2010)

Charles Darwin

Frederick Burkhardt et al (eds), *The Correspondence of Charles Darwin* (Cambridge: Cambridge University Press, 1985–)

Janet Browne, *Charles Darwin: A Biography*, 2 vols (New York: Alfred A. Knopf; London: Pimlico, 1995 and 2002)

Charles Darwin, *Evolutionary Writings*, ed. James A. Secord (Oxford and New York: Oxford University Press, 2008)

Adrian Desmond and James Moore, *Darwin* (London: Michael Joseph, 1991)

Sandra Herbert, *Charles Darwin, Geologist* (Ithaca: Cornell University Press, 2005)

Gregor Mendel

Arthur D. Darbishire, *Breeding and the Mendelian Discovery* (London: Cassell, 1911)

Hugo Iltis, *Life of Mendel* (London: George Allen & Unwin, 1932)

VítÐzslav Orel, *Gregor Mendel: The First Geneticist* (Oxford and New York: Oxford University Press, 1996)

Robert Olby, *Origins of Mendelism*, 2nd edn (Chicago: University of Chicago Press, 1975)

Roger J. Wood and Vítezslav Orel, *Genetic Prehistory in Selective Breeding: A Prelude to Mendel* (Oxford and New York: Oxford University Press, 2001)

Jan Purkinje

Henry J. John, *Jan Evangelista Purkyně: Czech Scientist and Patriot 1787–1869* (Philadelphia: American Philosophical Society, 1959)

Vladislav Kruta, *J. E. Purkyně (1787–1869), Physiologist: A Short Account of his Contributions to the Progress of Physiology with a Bibliography of his Works* (Prague: Czechoslovak Academy of Sciences, 1969)

Jan E. Purkinje, *Opera Omnia*, 12 vols (Prague: Society of Czech Physicians, 1918–1973)

Nicholas J. Wade and Josef Brožek, *Purkinje's Vision: The Dawning of Neuroscience* (Mahwah, NJ: Lawrence Erlbaum, 2001)

Santiago Ramón y Cajal

Santiago Ramón y Cajal, *Recollections of My Life*, trans. E. H. Craigie with the assistance of J. Cano (Cambridge, MA: MIT Press, 1989)

————, *Textura del Sistema Nervioso del Hombre y los Vertebrados* (1894–1904), English trans. *Histology of the Nervous System of Man and the Vertebrates*, trans. N. Swanson and L. W. Swanson (Oxford and New York: Oxford University Press, 1994)

————, 'The structure and connexions of neurons', Nobel Lecture, 12 December 1906, in *Nobel Lectures: Physiology or Medicine 1901–1921* (New York: Elsevier, 1967)

Benjamin Ehrlich, *The Brain in Search of Itself: Santiago Ramón y Cajal and the Story of the Neuron* (New York: Farrar, Straus and Giroux, 2022)

Marcus Jacobson, *Foundations of Neuroscience* (New York: Plenum, 1993)

F. Reinoso-Suarez, 'Cajal: a modern insight in neuroscience', in S. Grisolia, C. Guerri, F. Samson, S. Norton and F. Reinoso-Suarez (eds), *Ramón y Cajal's Contribution to the Neurosciences* (New York: Elsevier, 1983)

Francis Crick and James Watson

Francis Crick, *What Mad Pursuit: A Personal View of Scientific Discovery* (New York: Basic Books, 1988)

John Inglis, Joseph Sambrook and Jan Witkowski (eds), *Inspiring Science: Jim Watson and the Age of DNA* (Cold Spring Harbor, NY: Cold Spring Harbor Laboratory Press, 2003)

Victor K. McElheny, *Watson and DNA: Making a Scientific Revolution* (Cambridge, MA: Perseus, 2003)

Robert Olby, *Francis Crick: Hunter of Life's Secrets* (Cold Spring Harbor, NY: Cold Spring Harbor Laboratory Press, 2009)

Matt Ridley, *Francis Crick: Discoverer of the Genetic Code* (New York: HarperCollins, 2006)

James Watson, *The Double Helix: A Personal Account of the Discovery of the Structure of DNA*, ed. Gunther S. Stent (New York and London: W. W. Norton, 1980)

BODY AND MIND
Andreas Vesalius

Charles D. O'Malley, *Andreas Vesalius of Brussels, 1514–1564* (Berkeley: University of California Press, 1964)

K. B. Roberts and J. D. W. Tomlinson, *The Fabric of the Body: European Traditions of Anatomical Illustration* (Oxford: Clarendon Press, 1992)

Jonathan Sawday, *The Body Emblazoned: Dissection and the Human Body in Renaissance Culture* (London and New York: Routledge, 1995)

http://vesalius.northwestern.edu/flash.html

William Harvey

Leon G. Fine, *The Young Harvey* (London: Royal College of Physicians, 2004)

Robert G. Frank, *Harvey and the Oxford Physiologists* (Berkeley: University of California Press, 1980)

Geoffrey Keynes, *The Life of William Harvey* (Oxford: Clarendon Press, 1966)

Robert Willis, *The Works of William Harvey* (New York and London: Johnson Reprint Corporation, 1965)

Louis Pasteur

René Dubos, *Pasteur and Modern Science* (Madison, WI: Science Tech Publishers, 1988)

Gerald L. Geison, *The Private Science of Louis Pasteur* (Princeton: Princeton University Press, 1995)

Louise Robbins, *Louis Pasteur and the Hidden World of Microbes* (Oxford and New York: Oxford University Press, 2001)

René Vallery-Radot, *The Life of Pasteur* (London: Constable, 1906)

Francis Galton

Martin Brookes, *Extreme Measures: The Dark Visions and Bright Ideas of Francis Galton* (London: Bloomsbury, 2004)

Michael Bulmer, *Francis Galton: Pioneer of Heredity and Biometry* (Baltimore: Johns Hopkins University Press, 2001)

Derek W. Forrest, *Francis Galton: Victorian Genius* (New York: Taplinger, 1974)

Nicholas W. Gillham, *A Life of Sir Francis Galton: From African Exploration to the Birth of Eugenics* (Oxford and New York: Oxford University Press, 2001)

Sigmund Freud

Ernst Freud, Lucie Freud and Ilse Grubrich-Simitis, *Sigmund Freud: His Life in Words and Pictures* (London: Penguin, 1985)

Sigmund Freud, 'An outline of psychoanalysis' (1939) *Standard Edition of the Complete Psychological Works of Sigmund Freud*, ed. and trans. James Strachey, vol. 23 (London: Hogarth Press, 1953–74)

———, 'The psychopathology of everyday life' (1901), *Standard Edition of the Complete Psychological Works of Sigmund Freud*, ed. and trans. James Strachey, vol. 6 (London: Hogarth Press, 1953–74)

Ernest Jones, *Sigmund Freud: Life and Work* (London: Hogarth Press, 1953)

Alan Turing

B. Jack Copeland (ed.), *Alan Turing's Automatic Computing Engine* (Oxford and New York: Oxford University Press, 2005)

——— (ed.), *The Essential Turing* (Oxford: Clarendon Press, 2004)

——— et al, *Colossus: The Secrets of Bletchley Park's Codebreaking Computers* (Oxford and New York: Oxford University Press, 2006)

Andrew Hodges, *Alan Turing: The Enigma* (London: Burnett/ Hutchinson, 1983)

Charles Petzold, *The Annotated Turing: A Guided Tour through Alan Turing's Historic Paper on Computability and the Turing Machine* (Chichester: John Wiley, 2008)

Sara Turing, *Alan M. Turing* (Cambridge: W. Heffer, 1959)

John von Neumann

William Aspray, *John von Neumann and the Origins of Modern Computing* (Cambridge, MA: MIT Press, 1990)

Ananyo Bhattacharya, *The Man from the Future: The Visionary Life of John von Neumann* (London: Allen Lane, 2021)

Herman H. Goldstine, *The Computer from Pascal to von Neumann* (Princeton: Princeton University Press, 1972)

Steve J. Heim, *John von Neumann and Norbert Weiner: From Mathematics to the Technologies of Life and Death* (Cambridge, MA: MIT Press, 1980)

Norman Macrae, *John von Neumann* (New York: Pantheon, 1992)

A. H. Taub (ed.), *Collected Works of John von Neumann*, 6 vols (London: Pergamon Press, 1961–63)

Louis Leakey and Mary Leakey

Louis S. B. Leakey, *White African* (London: Hodder & Stoughton, 1937)

Mary D. Leakey, *Disclosing the Past* (London: Weidenfeld and Nicolson, 1984)

Virginia Morell, *Ancestral Passions: The Leakey Family and the Quest for Humankind's Beginnings* (New York: Simon & Schuster, 1995)

John Reader, *Missing Links* (Boston: Little, Brown, 1981)

POSTSCRIPT
Science and Scientists in our Time

American Chemical Society, 'The Keeling Curve', 2015: https://www.acs.org/content/acs/en/education/whatischemistry/landmarks/keeling-curve.html

Tim Berners-Lee with Mark Fischetti, *Weaving the Web: The Original Design and Ultimate Destiny of the World Wide Web* (New York: HarperOne, 1999)

Frank Close, *Elusive: How Peter Higgs Solved the Mystery of Mass* (London: Allen Lane, 2022)

J. Craig Venter, *A Life Decoded: My Genome, My Life* (London: Allen Lane, 2007)

SOURCES OF ILLUSTRATIONS

INDEX

Figures in **bold** refer to main entries; *italic* refer to plate illustrations